Everything is Design

The Hidden Ethics of Our Objects and Public Spaces

Alan J. Reid, PhD
Johns Hopkins University
Coastal Carolina University

Series editor
Graham Cairns
AMPS (Architecture, Media, Politics, Society)

The Interdisciplinary Built Environment

VERNON PRESS

www.vernonpress.com

In the Americas:
Vernon Press
1000 N West Street, Suite 1200
Wilmington, Delaware, 19801
United States

In the rest of the world:
Vernon Press
C/Sancti Espiritu 17,
Malaga, 29006
Spain

The Interdisciplinary Built Environment

Library of Congress Control Number: 2025933291

ISBN: 979-8-8819-0258-2

Table of Contents

List of Figures

List of Tables

Development of the Book

This book is an original manuscript. No funding was provided, and there are no conflicts of interest with the author.

About the author

Alan J. Reid, PhD is an Assistant Research Scientist in the Center for Research and Reform in Education (CRRE) at Johns Hopkins University where he primarily conducts mixed-methods research studies and design reviews on the efficacy of learning technologies. In addition, he is an Adjunct Teaching Faculty in the School of Education and teaches a doctoral-level course in the Diffusion of Technology Innovations. Dr. Reid is also an Associate Professor of Writing and Instructional Technologies at Coastal Carolina University, where he teaches graduate courses in research and writing and undergraduate courses in new media, digital culture and design. He received tenure in 2021.

His previous works include *The Smartphone Paradox* (2018) and *A Philosophy of Gun Violence* (2022), both of which are precursors to this book, which seeks to evaluate technological mediation through the lenses of ethical behavior and decision-making. Dr. Reid has given several public lectures and appeared at numerous conferences to discuss the central theme of these publications: Technology is non-neutral, which defines our identities, compels our behaviors, and affects our psychological and emotional wellbeing.

Preface

This book synthesizes several different disciplines as they relate to design, ethics, and the built environment. First, there is the socio-technological aspect of technology mediation. Our objects, according to philosophers like Don Ihde, Peter-Paul Verbeek, and Bruno Latour, mediate our experiences with the world around us. Through their designs (and, by extension, our perceived affordances), we largely comply with what our objects and spaces want us to do. At the micro-level, the phones in our pockets command our attentive processes in order to feed an attention economy. At the macro-level, urban planning and infrastructure can both promote inclusivity and foment violence. We are deeply intertwined with the objects and spaces that have been designed for us, sometimes deliberately and sometimes in unsuspecting ways.

Second, there is design ethics. Baked into every object, process, system, and environment are the remnants of the designer's morals, ethics, values, and biases. A popular example of this is hostile design, which weaponizes architectural elements of the built environment to deter and even prohibit undesired interactions between objects and people. (Take a walk around most urban areas, and you'll notice the ubiquity of spiked ledges, kinked rails, and divided benches that are explicitly designed to repel or exclude groups, often those who are marginalized and/or powerless). Ethics – to put it concisely – is mindful wellbeing. It is a set of standards that guides how we treat ourselves, one another, and the environment. When we design things for public use, we also communicate an ethical perspective. When we use things designed for us, we tacitly endorse their ethics.

The structure of this book follows a logical progression, which first defines ethical design (Chapter 1), provides frameworks, theories, and models for ethical decision-making (Chapter 2), and then presents individual chapters (Chapters 3-8), each one unique to a single area of our surrounding world (the built environment, nature, crime, animals, humans, and technology) along with thought-provoking case studies of real-world examples of ethical design and decision-making within that specific area. The final chapter (Chapter 9) discusses the future of design and how ethics will most certainly play a role in emerging technologies and environments. Last, the book proposes a checklist for the ethical evaluation of systems, products, and processes found in our shared objects and public spaces.

Importantly, this book seeks to cultivate mindfulness of the reader's interactions, providing them with a line of inquiry that questions areas of unethical design in their built environment and offers useful critiques and new solutions to these ethical dilemmas. We often have the power to reject those things that are irresponsibly designed and unethical in nature, and it is through this agency as users that we can demand better from designers, developers, and companies. It is imperative to understand our mediated relationships with the built and natural environments that surround us and the objects found within; this can help explain our behaviors and empower us to make ethical decisions that serve future generations.

This book is intended for all people who contemplate design. Whether you are a student pursuing a formal education in design school or you are simply interested in how the design of our shared objects and public spaces impacts our lives in implicit and explicit ways, this book is for you. If you have ever been frustrated with an object or a space that fails to deliver on its promise, this book is for you. If you feel that designers have a shared responsibility to their users and to the rest of society, this book is for you.

The chapters that follow are organized thematically. Each one centers on a general subject area that includes an abundance of examples, some of which can seem disconnected from one another, but that all share a common premise: Because we interact with design every day and in everything, we should take notice of the underlying ethics that are manifested in the shared objects and public spaces of our natural and built environments. In doing so, we become empowered as consumers, as citizens, and as agents of change. This book will help you become more mindful of your surroundings by realizing the intentional designs that impact our lives, and the ethics that are baked into them.

There are several assumptions under which this book operates. First, many of the concepts and examples that are provided assume that the reader has some degree of familiarity with them. For instance, you do not have to have ridden in a self-driving car to understand its relevance to the discussion of its ethics, but using this as an example does require that you possess an awareness of autonomous cars and how they might be used. For some examples, a brief description is provided to orient the reader, who may be totally unfamiliar with the concept. Second, this book assumes a great deal of privilege. Design is essential for survival in many cases and in many places around the world. Therefore, it must be recognized that we are in a highly privileged position to evaluate systems, products, and spaces for their ethics and not just their utility.

This book is not a comprehensive review of scholarly literature, nor does it lay claim to new scholarship in design studies. Instead, it is a collection of observations that interrogates our environment through the lens of ethical design. If you have ever wondered why something works the way it does, why it is not better, or whether we should be supporting it in the first place, then this book may be useful for you. And truth be told, it is very likely that you will have more questions than answers after reading this book. The aim of this book is to provoke a critical curiosity into the designs of our objects and spaces that envelope you and that were developed by some other human in some other time and place, who is imperfect, and whose ethics are baked into that design.

Chapter 1

What is ethical design?

As we 'make things work,' what kind of world are we making?[1]
Langdon Winner

This book is a long-form attempt to make the case for the ethical design of our objects and our shared spaces. Of course, this is predicated on having a clear and common understanding of two concepts, in particular, *ethics* and *design.* But articulating a universally accepted explanation of each is an exercise in futility. Conceptual ideas like philosophy, morals, values, good, and bad are highly contextual and bound by spatial and temporal constraints. And so, this book is not going to try to offer a definitive explanation for what *ethics* or *design* is; instead, this book will guide you through many different subject areas in which both ethics and design are collaborative forces that collide, resulting in ethical and unethical objects and spaces. The intention of this book is to spotlight examples where design lacks ethics and to celebrate others who promote ethical design. To do so, we must have a baseline for discussing *ethics* and *design.*

What does it mean to be *ethical*? Can the design of an everyday object, like a coffee maker, have *ethics*? The short answer is yes; a coffee maker can have ethics. Well, sort of. The coffee maker itself is a physical object that is a blend of molded plastics and various glass and metals that are formed into individual components – mainly a housing unit, heating element, water reservoir, decanter, and warming plate – which make up the object that we would recognize as a coffee maker. But behind every product, environment, process, and system is a human designer (or team of designers) who espouses certain beliefs, morals, values, and biases. These human elements infiltrate the design of the product in various ways that can influence a user's behaviors, have an impact on the environment, marginalize others, and more. The broader practice of *design ethics* ensures that things are designed, built, and implemented responsibly and with the collective good in mind.

As an example, let's consider Keurig, a U.S.-based manufacturer that pioneered the single-serve coffee maker. The Keurig coffee machine was launched in 1998 by four entrepreneurs who designed the single-serve brewing system along with its "K-Cup" pods for brewing individual cups of coffee. However, Keurig did not invent a single coffee pod (this was introduced by Nespresso in the 1980s), and many competitors have emerged since. Roughly half of all American households use a Keurig, and there are more than 25 million Keurig and K-Cup brewers in homes and offices across the United States. Keurig had made brewing a single-serve cup of coffee more convenient, but at what cost? For instance, in 2013, Keurig produced 30 billion K-Cup pods, and it is estimated that roughly 13 billion pods ended up in landfills that same year. Despite the messaging on the product label, K-Cup coffee pods are not entirely recyclable. It is not surprising, then, that in 2023, Keurig Green Mountain, Inc. settled a class action lawsuit, which contended that "because the majority of recycling facilities cannot segregate such small waste from the general waste stream, and because there is no market to recycle the Products, the recycling facilities send the Products to landfills." An online marketing campaign has since gone viral on social media with the hashtag #KillTheKCup, and one of the original founders, John Sylvan, said this about his invention of the Keurig: "I feel bad sometimes that I ever did it."[2] The company has since made efforts to manufacture K-Cups out of recyclable plastic, but it still requires the user to peel off the aluminum top, empty the grounds in the trash, and rinse the container before placing it in the recycling bin – an inconvenient practice that most users neglect to do since their primary motivation for using a K-Cup is convenience. *Consumer Reports* has concluded that "the best solution to coffee pod waste is to ditch pod coffee makers altogether and switch to other coffee-brewing methods."[3]

Certainly, the popularity of the K-Cup has been profitable for its shareholders, but we might ask whether the design of the single-serve coffee pod was *ethical* in the first place. The K-Cup did not necessarily introduce a new product; it merely provided consumers with a new delivery mechanism for the convenience of a single-serve cup of coffee, but in doing so, contributed to landfill waste by orders of magnitude. An ethical consumer might ask: Did the innovation of this product serve the collective good, or was it a short-sighted development with only the company's interest in mind that will have lasting impacts on the environment?

And what about the concept of *design*? How might we define this vague yet familiar term? As a matter of fact, design can be observed everywhere and in everything. On the macro level, design can be the layout of an entire city or neighborhood, or the rerouting of entire rivers to irrigate arid regions, or giant windmills and solar farms that clash with natural landscapes. On the micro level, design informs the phone in your pocket, the food from your last meal, and the chair that you are sitting in right now. Everything is designed, either by humans (the built environment) or naturally, through evolutionary selective processes (the natural world). A design is ethical when it considers the best interest of those affected by the design and implementation of the object, thing, or artifact. *Ethical design* is that set of standards and practices that guides how an object is shaped and subsequently affects its users as well as those in its periphery.

It is likely that you, dear reader, will not pursue a career in product design, architecture, urban planning, web design, terra-farming, or engineering; in fact, this book is not meant to be an instructional guidebook for these disciplines. Instead, the purpose of this book is to cultivate mindfulness of your interactions with the surrounding world and built environment, providing you with a line of inquiry that questions areas of unethical design and offers useful critiques and new solutions. Even though you are not the one who designed and built the object, you often have the power to reject those things that are irresponsibly designed and unethical in nature, and it is through this agency as users that we can demand better designers, developers, and companies. That said, this book is not just an argument for consumer responsibility. Sometimes, we simply do not have a choice whether we travel on a road that racially segregated a neighborhood, eat a meal that came from factory-farmed animals, or live in a country that deploys killer drones on foreign civilians. Our agency is limited, but at the very least, this book aims to create a mindfulness of how ethics inform the design of our everyday objects and spaces.

The chapters in this book cover a wide range of topics, from architecture and the natural environment to crime and safety to humans and technology, and you will be introduced to different frameworks and models that can be used to evaluate the ethical designs within each. You will be given many real-world cases to contemplate, and you will be asked to examine different scenarios through the lenses of *ethics* and *morals*. But first, we must understand the differences between the two.

Ethics v. morals

Although most people use the terms *ethics* and *morals* interchangeably, they do have different meanings. Importantly, when we discuss *ethics*, we are not saying that something is *right* or *wrong*, *good* or *bad*. Ethics and morals both connote a larger set of principles and beliefs that guide our behaviors, but acting ethically is not always the same as acting morally, and vice versa.

An example of conflicting morals and ethics is the Westboro Baptist Church (WBC), which has been classified as a hate group by the Southern Poverty Law Center and the Anti-Defamation League primarily because of its homophobic and anti-Semitic agenda and other extremist views on social and political issues. The church was founded in 1955 and has risen to notoriety because of its frequent protests at funerals for AIDS victims. Members of the WBC claim to have picketed more than 40,000 times, including at the World Trade Center Memorial on the anniversary of September 11th and at cemeteries during fallen soldiers' funerals, claiming that "God is punishing America for tolerating homosexuality."[4] A 2015 study surveyed members of the religious group, all of whom agreed with the statement that "God causes major tragedies to occur as a warning to sinners."[5] For members of the WBC, their reprehensible actions are perceived as being justified because their beliefs dictate that homosexuality is a moral abomination. Most people, however, would view these actions as being intolerant and generally unethical.

Table 1.1. Types of ethical lenses through which to view actions.

Ethics	**Morals**
Universal	Individual, Cultural, Societal
Derived from Latin, *ethos*, or "character."	Derived from Latin, *mores*, or "conventions."
Refers to guiding principles for the way things 'ought to be.' Ethics encourage us to behave in accordance with the well-being of ourselves and of others.	Often, faith-based, moral principles outline what is believed to be 'right and wrong.' This is contextualized within the era and the norms of cultural and social groups.
- EXAMPLE -	
The professional ethics of an attorney demands that criminally charged defendants receive a right to counsel and legal representation. In the United States judicial system, defendants are presumed innocent until proven guilty and promised the right to a fair trial, regardless of how overwhelming the evidence of guilt may be.	Murder is a morally reprehensible act, so an attorney might object to representing a criminal defendant on moral grounds. If the lawyer lost a loved one to gun violence, they may wish to abstain from representing someone accused of homicide-by-gun.

Simply put, *morals* are subject to individualistic and cultural beliefs, whereas *ethics* transcend the individual, community, and society. An article published in 1928 in the *International Journal of Ethics* makes the following distinction between the two: "Ethics is not an attempt to regulate conduct. Attempts to do so are moralities; and ethics is not morality."[6] The table above illustrates the differences between ethics and morals.

As demonstrated above in the WBC example, acting morally is dependent upon the context. A *moral absolutist* will argue that there is a fixed set of rules that dictate whether something is right or wrong. A *moral relativist* will argue that good and bad can change depending on the circumstance. According to The Ethics Centre, there are "many valid moral systems," and the "key question for a moral relativist is what morality ought to be relative to."[7] The Westboro Baptist Church espouses a moral code that is ethically dishonest because it promotes destructive behavior towards others, even though they feel morally obliged to do so.

To be *ethical* does not mean to be morally *perfect*. In the context of design, being ethical simply means that the main intent behind the use of a product is good (which, admittedly, can be a relative term) and that the making process is not knowingly or deliberately harmful to others or to the built or natural environment. According to Peter-Paul Verbeek, a Professor of Philosophy and Ethics:

> Even when designers do not explicitly reflect morally on their work, the artifacts they design will inevitably play mediating roles in people's actions and experiences, helping to shape moral actions and decisions and the quality of people's lives. Technology design, therefore, appears to be an inherently moral activity.[8]

Ethics and morals are not always mutually agreeable; a technological product, environment, process, or system can be ethically good but violate a moral code. For instance, roughly 5 million Americans require blood transfusions every year. This medical breakthrough was first performed in 1667 and is now considered routine. The medical procedure has been lifesaving for many people (including my own father), but for the Christian denomination of Jehovah's Witnesses, receiving blood violates their religious beliefs and, by extension, their moral code. But suppose an individual of Jehovah's Witness faith requires a life-saving blood transfusion, and declining this care would mean certain death. Furthermore, we might also consider whether the doctor or the hospital is obligated to treat the individual, citing their own ethical responsibility to the Hippocratic Oath to "do no harm." While a Jehovah's Witness might be morally opposed to blood transfusion, the doctor might be ethically obligated to save a patient's life by any means necessary. The result is a stalemate between *ethics* and *morals.*

The two examples I just provided – the WBC and Jehovah's Witnesses – highlight the complex relationship between religious belief and morality. Often, it is argued that one cannot be moral without a religious foundation. That is, without religion, there would be no moral framework to guide society to do the right thing. But this is simply not the case; people make moral and ethical decisions regardless of their religious beliefs (or absence thereof) all the time. To show this, researchers at Harvard University developed the Morality Sense Test (MST)[9] consisting of the three scenarios presented below. Participants were asked to fill in the blanks using the phrases "obligatory," "permissible," or "forbidden," and each phrase could only be used once. Go ahead and try this exercise:

Table 1.2. Morality Sense Test

1	*A runaway boxcar is about to run over five people walking on the tracks. A railroad worker is standing next to a switch that can turn the boxcar onto a sidetrack, killing one person, but allowing the five to survive. Flipping the switch is ______.*
2	*You pass by a small child drowning in a shallow pond, and you are the only one around. If you pick up the child, she will survive, and your pants will be ruined. Picking up the child is ______.*
3	*Five people have just been rushed into a hospital in critical condition, each requiring an organ to survive. There is not enough time to request organs from outside the hospital, but there is a healthy person in the hospital's waiting room. If the surgeon takes this person's organs, he will die, but the five in critical care will survive. Taking the healthy person's organs is ______.*

If you identified Scenario #1 as permissible, Scenario #2 as obligatory, and Scenario #3 as forbidden, you would be in agreement with more than 90% of respondents who took the test. Importantly, researchers found that a person's religious background did not predict or correlate with their responses, presumably showing that we all have an ethical and moral baseline by which we live, regardless of our individual faiths (or lack thereof). Or, as the authors put it, we are "endowed with a moral faculty that guides our intuitive judgments of right and wrong. These intuitions reflect the outcome of millions of years in which our ancestors have lived as social mammals and are part of our common inheritance."[10] Religion and morality are often coupled together, but it turns out that we all have a similar barometer for judgment.

We shouldn't think of ethics and morals in terms of being either *good* or *bad,* or even being at odds with one another. Oftentimes, things are both. Consider your smartphone, for example. This powerful mobile device can connect us to loved ones, provide us with useful news and information, and be used to capture memorable photos and videos. But with these benefits also come drawbacks; the smartphone exploits our attentive processes, promotes conspiracies and misinformation, and commodifies our personal data. So, relatively speaking, the smartphone is both good and bad, depending on your perspective. If a parent knows that social media usage is linked to poorer mental health in teenage girls, then the decision to buy a smartphone for their pre-teen daughter becomes a sort of cost-benefit ethical decision to be made, whether parents are mindful of this or not. Similarly, smartphones are neither ethical nor unethical devices; they are both, depending on context. A really well-designed social media app that is designed to engage young girls in social comparison to one another is good for the app stakeholders but not for the users. We should not confuse ethical positivity with design quality.

The goodness or badness of objects can shift over time. For example, in the mid-1990s, the pharmaceutical company Purdue Pharma manufactured the drug OxyContin, which was approved by the Federal Drug Administration (FDA) as a pain reliever that was legitimately prescribed by healthcare professionals to help manage pain after moderate to serious injuries and routine surgeries. While this seems like a good-faith advancement in medicine initially, the over-prescription of this drug and similar opioid products led to an epidemic. The Congressional Budget Office (CBO) asserts that the first opioid crisis began in 1996, and a fourth wave of opioid-related deaths began in 2019. The Centers for Disease Control and Prevention (CDC) reports that there were nearly 92,000 drug overdose deaths in 2020, 75% of which were attributed to opioids. The CDC estimates that the opioid crisis costs the U.S. more than $1 trillion every year. In 2022, the Sackler family – who owns Purdue Pharma – settled a $6 billion lawsuit in which they admitted to criminal wrongdoing in their aggressive marketing of OxyContin as a safe prescription medicine, downplaying its addictive properties. It is unlikely that Purdue Pharma originally had evil intentions to unleash an opioid epidemic on society, but once the drug's devastating effects had become apparent, the ethical behavior would have been to limit drug production or, at the very least, make an earnest effort to educate medical professionals and the public about the dangers of OxyContin instead of pushing it even harder.

The objects in our lives mediate our experiences with the world around us in explicit and implicit ways. In Verbeek's book *What Things Do*, he says that "Design ethics requires that artifacts be treated as members of the moral community" and that they shape our behaviors "regardless of whether this is done consciously and intentionally or not."[11] The latter part of this quote is especially significant: *consciously and intentionally or not.* It is imperative that the evaluation of ethics is ongoing; it does not merely exist within the design process since objects are continually evolving, finding new uses, and realizing new effects, whether they were conscious and intentional or not. When designers and developers neglect their duty to pursue ethical design for well-being, the responsibility often falls onto the individual users, which is a breach of the implied contract that we have with one another as citizens, consumers, and fellow human beings.

The social contract

At the heart of this text is the underlying assumption that we should all behave ethically. But why *should* we behave ethically? It's not enough just to know what ethical behavior looks like; we also must be convinced that it is worth pursuing. For instance, a capitalist might be less concerned with behaving ethically and more interested in the bottom line. A criminal might be driven by self-interests and not the greater good. A politician might be willing to put their party over their country, even if it means disenfranchising millions of people. Perhaps the question is not whether we ought to behave ethically but whether we are naturally included to do so. Philosophers have long considered whether humans possess this innate ability to self-regulate their behaviors, or whether this must be governed explicitly by political and legal structures. Thomas Hobbes (1588-1679) was a seventeenth-century English philosopher who argued that individuals behave according to their own interests, which he labeled our "state of nature." Only through something called the *social contract* will individuals harness these self-interests for the greater good of society.

The social contract theory was promoted throughout the seventeenth and eighteenth centuries by public intellectuals like John Locke (1632-1704),

Immanuel Kant (1724-1804), and perhaps most notably, Jean-Jacques Rousseau (1712-1778). Rousseau's book, *The Social Contract*, was published in 1762 in France and served as the foundational work for many modern political philosophers. Social contract theory is complex, but it essentially states that by virtue of living in a society, we all have a vested interest in (and thereby an obligation to) behaving ethically and morally for the benefit of others. Sometimes, this is enforced explicitly by a governing body (e.g., speed limits are set by local and state legislatures to promote responsible driving. Failure to observe posted speed limits may result in fines and even a suspension of your license). Other times, our behaviors are deterred by an ethical arrangement that is implied with the rest of society (e.g., walking into a grocery store and taking someone's shopping cart that is full of groceries is not illegal, but it violates our unwritten social code of conduct). The purpose of the social contract is to ensure that everyone is treated fairly and with respect, and this concept extends to the ethical design of the products, environments, processes, and systems that we interact with every day.

The *common good* is a phrase often associated with various political, economic, and social philosophies, and is the footing for our modern-day social contract with one another. Essentially, the common good principle advocates for everyone to "live well [and] to experience the fullness of human life."[12] In his book *The Common Good*, Robert Reich writes that "The common good consists of our shared values about what we owe one another as citizens who are bound together in the same society – the norms we voluntarily abide by, and the ideals we seek to achieve."[13] The common good is facilitated by our things – products, environments, processes, and systems – which can be designed ethically to promote, or at the very least, nudge us towards making decisions that support the collective wellbeing. Like ethics, the common good is universal; it is not subject to societal norms, cultural practices, or the accepted behaviors during a specific era. The common good always governs over all of us, but it is up to us to maintain it and fight to preserve it, and sometimes, this means direct enforcement.

In the 1960s and 1970s, there was a wave of legislative efforts to require motorcyclists to wear helmets while riding. Research consistently showed that wearing a helmet significantly reduces the risk of injury and fatality in motorcycle accidents, and it was this rationale that informed state-mandated motorcycle helmet laws (although three states – Illinois, Iowa, and New Hampshire – still do not have motorcycle helmet laws at the time of this writing).[14] Despite the evidence that helmets make motorcycle riding safer, the motorcycle helmet laws were challenged repeatedly in court, citing a rider's constitutional right to privacy. Unarguably, the 9th Amendment of the United States Constitution guarantees an individual's inherent right to privacy, but a 1986 report co-sponsored by the University of Virginia and the Virginia Highway and Transportation Research Council refuted this constitutional argument with the "missile hazard theory", which proposed that a motorcyclist not wearing a helmet is unprotected, and "vulnerable to many roadside hazards… which might strike the biker, causing him to lose control." Therefore, "an out-of-control cyclist is a threat to everyone else on the highway…[and] helmet laws are a reasonable means of serving a legitimate public end, the safety of all those who travel on the streets and highways."[15] The authors cited philosopher John Stuart Mill, quoting "Sic utere tuo et alienum non laedas" or "So use your own that you do not injure that of another." In this case, the common good outweighs the individual's decision to wear a helmet while riding a motorcycle.

The motorcycle helmet laws would serve as the template for requiring seat belts in automobiles. In the 1980s, the federal government instituted a mandate for all U.S.-based auto manufacturers to build seatbelts into their vehicle designs after it was concluded that seatbelts minimize the risk of serious injury and fatality in car accidents.[16] In fact, all but one state (New Hampshire) legally requires occupants in the front seat to wear a seat belt. Additionally, modern automobiles are federally required to be equipped with seat belt reminders after it was found that this design feature effectively increases seat belt use. Although research clearly demonstrates that wearing a seat belt drastically reduces the risk of injury or death, many opposed seat belt restraint laws initially.

A 1987 op-ed published in *The Chicago Tribune* by William J. Holdorf likened the state-mandated seat belt laws to the 1920s Prohibition era, arguing that "saving freedom is more important than trying to regulate lives through regulation." Holdorf concluded that "Safety imposed is freedom lost."[17] Indeed, the 'freedom' argument is a common rhetorical move for those who oppose any sort of behavioral regulation (e.g., unfettered access to guns and open and concealed gun-carrying, the refusal to become vaccinated or to wear masks in public spaces during a pandemic, or even the decision to drink and drive). But importantly, these regulations are less of an Orwellian attempt at legislating behavior and more about the ethical obligation to protect others who are tangentially affected by these "personal" decisions.

There is a delicate balance between supporting the common good and preserving individual freedoms. Although research shows that smoking cigarettes and exposure to secondhand smoke have adverse effects on one's health, smoking cigarettes is not an illegal activity for those over the age of 18. Adults have the right to choose whether they engage in this behavior; this is indisputable. However, when a woman smokes during pregnancy, her personal choice also may affect her unborn child. The Center for Disease Control and Prevention (CDC) reports that smoking while pregnant increases the likelihood of preterm delivery, lower birth weights, and birth defects. In fact, babies with mothers who smoke during pregnancy or who expose their babies to dangerous amounts of secondhand smoke are three times more likely to die from sudden infant death syndrome (SIDS).[18] It can be argued that a mother has a moral and ethical duty to protect her child, yet smoking while pregnant imposes undue risk on the child. Indeed, the law distinguishes between *absolute rights* and *prima facie rights*, where the former cannot be denied to any individual, and the latter can be outweighed by other considerations. As you can see, personal freedoms and the rights of others often clash.

During a State of the Union Address in 1941, just prior to American involvement in World War II, President Roosevelt outlined his Four Freedoms, which would later inform the basis for the United Nations and its central document: the Universal Declaration of Human Rights. These freedoms are as follows: the freedom of speech, the freedom of worship, the freedom from want, and the freedom from fear. The first two are clear and obvious references to the First Amendment of the U.S. Constitution, which guarantees the freedom ***of*** these individual rights. The other two, however, reference the freedom ***from*** something that could be imposed on the individual by others. Roosevelt's interpretation of the "freedom from want" referred to "economic understandings which will secure to every nation everywhere a healthy peacetime life for its inhabitants" and for the "freedom from fear," the "world-wide reduction of armaments to such a point…that no nation anywhere will be in a position to commit an act of physical aggression."[19] These are lofty standards for the world stage, but we might also apply these freedoms to our built environment and to the objects within. For example, we reserve the right to co-exist in spaces and use objects that are not harmful to us. Regulatory efforts ensure that the elevator we use is in compliance with strict safety standards and that the paint that we buy at the home improvement store is lead-free. As co-signers of the social contract, consumers must be able to trust that companies, designers, developers, and advertisers behave in ethical ways that benefit the common good of society and not just shareholders and their bottom line.

In the book *Philosophy and Design*, Peter-Paul Verbeek argues that our technological artifacts, materials, and environments are hopelessly entangled with our freedoms: "The concept of freedom presupposes a form of sovereignty with respect to technology that human beings simply no longer possess."[20] Our technologies – and the mediated experiences that they afford – both grant and impinge on our freedom, often without our consent (and sometimes without our knowledge). Since 2007, there have been more than 1,600 lawsuits filed in international, state, and local courts that have litigated the effects of climate change, many of which have been brought by young people under the age of 18.[21] The legal argument is that younger generations are entitled to a clean environment, and that corporations that delay climate change solutions are in breach of state, federal, and international environmental law. For instance, the first-ever climate trial in the United States (*Held v. State of Montana*) was brought in 2023 by 16 children represented by Our Children's Trust, who alleged that the state's continued pursuit of fossil fuel extraction and burning hastens the climate crisis and subsequently violates the state constitution, which reads: "All persons are born free and have certain inalienable rights. They include the right to a clean and healthful environment" (Article II, Section 3).[22] Certainly, preserving natural resources (or, at the very least, not actively polluting them) is a basic human right and the ultimate example of the social contract by which we live.

When things are designed for public use – whether it be a subway system or a social media app – we might consider how it meets (or doesn't meet) a baseline of ethical standards. The Ethical Hierarchy of Needs (shown below) is a framework articulated by Aral Balkan and Laura Kalbag, founders of Ind.ie, a "non-profit striving for social justice" that aims to "design and build ethical technology that respects human rights, effort, and experience."[23]

The triangular graphic invokes Maslow's Hierarchy of Needs and demonstrates a tiered approach to the ethical design of technology. Just as with Maslow's hierarchy, each level can only be met when the level below it has been satisfied. Thus, technology is not a "delightful" experience without functionality, convenience, and reliability, nor can it be those things without first respecting basic human rights. The top tier of human experience is only attained when technology "puts a smile on your face and makes your life better."[24] Summarily, ethical design is concerned not only with how something makes you feel, but how it impacts others.

Figure 1.1. The Ethical Hierarchy of Needs

Renowned philosopher Langdon Winner titled his 1986 essay, "Do Artifacts Have Politics?", and 28 years later, Peter-Paul Verbeek similarly inquired, "Do Artifacts Have Morality?" The answer to both questions, according to the philosophers, is yes, technological artifacts, objects, spaces, and our built environments have both politics and moralities, as well as ethics. In a public lecture on the Ethical Hierarchy of Needs, co-founder Aral Balkan says, "If you think about it, experiences are all we have in life, and then we die. Whether those experiences are with people or with objects, they are like the grains of sand in the hourglass of your life."[25] Verbeek reinforces this idea that ethical design directly relates to our quality of life, saying: "Ethics, then, was about excellence in living, or mastering the art of living. In a technological culture, an ethics of the good life is about developing forms of excellence in living with technology."[26] And Langdon Winner wrote, "[W]e repeatedly enter into a series of social contracts, the terms of which are revealed only after the signing."[27] This is precisely why ethical design matters; it not only honors the social contract that is the fabric of our society, but it demands that we address the non-neutral makeup of our objects, processes, and systems that lie within.

Value-sensitive design

So far, we've discussed ethics and morals and why it is important to account for them in design. But things also are value-laden. Unlike *ethics*, which strives for the common good, *values* depend on the individual and affect our judgments. Values reflect one's beliefs that guide their choices and behaviors, and like morals, they are highly contextual. Let's return to the example of the Keurig coffee machine. A quick Google search reveals that during the federal election cycle of 2019-20, the Political Action Committee (PAC) for Keurig Dr Pepper (the parent company of Keurig) donated $15K to Democrats and $41K to Republicans. These lopsided financial contributions might suggest a general favorability towards Republican candidates over their Democratic opponents and, by extension, perhaps broader support for conservative "values." If you identify as a Democrat, you might be less than thrilled to hear this about your favorite coffee maker; conversely, if you identify as a Republican, you might feel even more aligned with the Keurig brand. This is the case with every object you've used or any built environment that you've encountered in your entire life. Designers, developers, manufacturers, builders, and distributors all possess values – some of which you might vehemently oppose – and so making purchasing decisions based on this alone seems futile. Politics is infused into every object, brand, and company, even for the most seemingly benign things.

In 2017, the Unite the Right rally in Charlottesville, VA, featured a large gathering of white nationalists who marched in solidarity to protest the removal of a Confederate statue from a public park. The protestors carried tiki torches – the bamboo torch most often associated with luaus, beach parties, and backyard barbecues – which communicated a strong symbolism; Adolf Hitler wrote in *Mein Kampf* that racial purity is "the fuel for the torch of human culture." And as protestors marched throughout Charlottesville with tiki torches in hand, they were heard chanting racist slogans like "Jews will not replace us," "Blood and soil," and "The South will rise again." Despite a swift response from the company CEO denouncing the rally, the Tiki torch had been unfairly implicated in the white supremacist starter kit, and risk management consultants correctly predicted that it would take years for the company to distance itself from the Charlottesville rally that resulted in the deaths of multiple people. Indeed, the tiki torch continues to invoke an undercurrent of racism. In 2021, a group of men appeared at a campaign stop for the Republican candidate for Virginia governor, Glenn Youngkin, prominently displaying their tiki torches as a callback to the unfortunate events that took place in 2017.

Through no actions of their own, the TIKI® brand has suffered financial loss and will be forever connoted with white nationalism, yet these hateful political values were neither designed into the object itself nor espoused by anyone connected to it. These values were grafted onto the object. Certainly, if we dig deep enough, every object that we encounter in our daily lives will have something embedded in its past that we find disagreeable or even abhorrent. Perhaps the manufacturer of the product was involved in human rights violations (Volkswagen, Hugo Boss, and Adidas have roots in Nazi Germany), or the brand has supported a social justice cause that angers its consumer base (Anheuser-Busch faced a boycott after partnering with a transgender activist, Dylan Mulvaney), or the business operates under strict religious principles (Chick-fil-A has been closed on Sundays since 1946 to observe the Sabbath). There are countless examples of how brand activism and brand politics have been shown to influence consumer attitudes and behaviors toward that brand. While the *values* of an object (or of its founder, CEO, or marketing team) are not the same as the *ethical design* of that object, neither is neutral.

Designers have come to understand the importance of value sensitivity and awareness, and in the 1990s, the concept of *value-sensitive design* (VSD) was born out of computing technologies. Batya Friedman, a Professor of Computer Science, is credited as being the one to first articulate VSD, which she explains in the following way:

> Value sensitive design seeks to guide the shape of being with technology. It positions researchers, designers, engineers, policy makers, and anyone working at the intersection of technology and society to make insightful investigations into technological innovation in ways that foreground the well-being of human beings and the natural world.[28]

Specifically, Friedman defines values as being "what is important to people in their lives, with a focus on ethics and morality."[29] But this can be very problematic for one obvious reason. According to an article published in *Ethics and Information Technology*, VSD does not commit to any particular ethical theory or framework; rather, a major critique of VSD is that it "shows *that* people hold certain values, [but] it is not able to say anything about whether people *should* hold certain values."[30] In other words, designing something nefarious or hateful could satisfy the values of a particular group of stakeholders but, in doing so, could violate basic ethical standards and negatively impact others. This suggests that VSD sometimes overlaps with but operates entirely independently from ethical design.

Since its inception in the mid-1990s, VSD has expanded into nearly all fields and levels of the human experience (ranging from the individual to global experiences). One of the most common methods of applying VSD is through stakeholder analysis. The purpose of this analysis is to identify two groups, specifically: ***direct*** stakeholders (those who use the technology or object) and ***indirect*** stakeholders (those who will be affected by others using the technology or object) and the values that they support. Often, the interests of these two groups will conflict. So, philosopher Ibo van de Poel has established a values hierarchy[31] that takes the values of stakeholders and translates them into actual design requirements for the product, environment, process, or system:

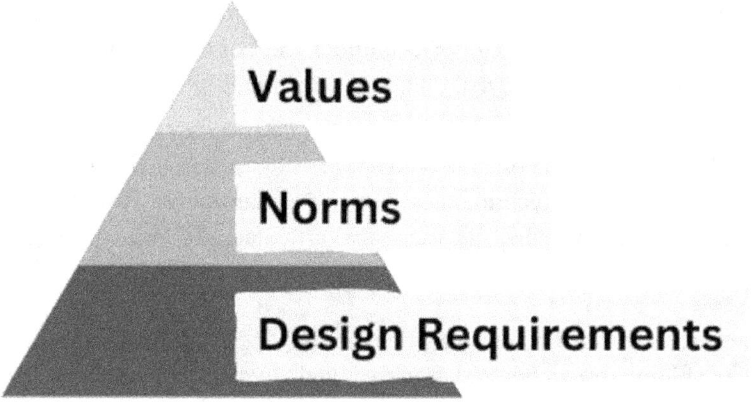

Figure 1.2. The Values Hierarchy

Using a top-down approach, the hierarchy moves from the abstract to the concrete. The designer first identifies a value – e.g., animal welfare, clean energy sources, organic farming, nuclear disarmament – deconstructs this into specific norms (what the design should be able to do), and then articulates design requirements to be embodied in the technology.

Urban-built environments often include explicit architectural and object design features to discourage or prevent unwanted behaviors (this is discussed in much more detail in Chapter 3). Let's use the example of the divided bench, an object commonly used to represent *hostile design* (sometimes referred to as defensive design, hostile architecture, and anti-homeless architecture). Most notable in the design is the raised arm that compartmentalizes the bench into seats. In doing so, the arms both define a space for an individual to sit down and prevent someone from lying down on the bench.

Figure 1.3. A divided bench in Lower Manhattan, New York City (Author photo)

New York City – where the above photo was taken – has experienced surging levels of homelessness in recent years, rivaling those of the Great Depression.[32] Yet divided benches, like the one pictured, are ubiquitous throughout the city and effectively deter unhoused people from sleeping on these public structures. Building a bench for the public suggests a welcoming environment; subdividing that bench to deter the unhoused suggests we are only welcoming *for some* and not others. If we apply van de Poel's value hierarchy, we might conclude the following:

> **Values**: The business or city wishes to provide a seating option for the public.
> **Norms**: The unhoused population should be prevented from sleeping here.
> **Design Requirements**: Install dividers or other deterrents to achieve the norm.

At this point, you might be wondering: What's wrong with a business or a city wanting to keep the unhoused from sleeping in public spaces? Isn't it their right (and freedom) to do so? The answer is yes – the proprietor of the bench does have the right to exclude whomever from their bench. But this text is concerned more with how this bench leverages design practices as enforcement of values. In fact, we can learn a lot about a society by how it utilizes its public designs. Kathy Madden and Fred Kent are the founders of the Social Life Project, whose mission is to "incite a renaissance of community connection in public spaces."[33] Their website highlights examples of well-designed environments that contribute to social positivity rather than defend from unwanted behaviors. They argue that public seating symbolizes more than a place to sit:

> Benches and seating are not just objects; they are mirrors to our social behavior. The ways people interact with seating can reflect their personalities, their sense of belonging and security in public spaces, and their openness to other people. To look closely at how people use something as simple as a bench is to shed light on the ways that we interact; the intricacies of how humans share spaces, even when it is time to sit down for a rest.[34]

The divided bench is a proxy for our negligence towards the unhoused. In his book *Callous Objects: Designs Against the Homeless*, Robert Rosenberger points out that the designs of our public spaces convey our ethics. Divided benches and other, more harmful designs like spiked ledges and acoustic deterrents "test our moral intuitions" and force us to address our "technological guilt."[35] Perhaps we should spend less time designing public spaces that exclude marginalized groups of people and instead focus our efforts on how to design for the care and well-being of those groups who are vulnerable, at-risk, and marginalized.

Figure 1.4. An Archisuit overcomes a hostile design (Source: Sarah Ross)

Sarah Ross is an artist who developed "Archisuits," which are body suits designed to overcome the design limitations of divided benches. Though the modified jogging suits are more of a social commentary piece than a functional consumer product, they were designed to "point at architecture as an arm of the law, a form that uses the built environment to police and control raced, classed, and gendered bodies."[36] These wearable suits make an unpleasant design a pleasant experience and, in doing so, demonstrate nonviolent resistance.

While the Archisuit is not a practical piece of clothing, it succeeds in raising a very important point: *Ethics informs design.* Every engineered product, environment, process, or system possesses humanistic qualities that are communicated to stakeholders. Langdon Winner wrote, "Through technological creation and many other ways as well, we make a world for each other to live in. Much more than we have acknowledged in the past, we must admit our responsibility for what we are making."[37] We must not only consider what things will do *for* others, but what they will do *to* others as well. We must acknowledge both *direct* and *indirect* stakeholders. And we must constantly reflect on those things and the ethics, morals, and values that they convey.

Case study: Public monuments

An essay by W.J.T. Mitchell titled "The Violence of Public Art," in *Beauty is Nowhere: Ethical Issues in Art and Design,* discusses how designing public monuments requires a complicated and delicate balance of functional design, commemoration, and timelessness. He uses the example of how the Vietnam Veterans' Memorial (VVM) in Washington, D.C., is actually a "cunning violation and inversion of monumental conventions."[38] Rather than depict a famous figure or scene, the monument is a land art-inspired granite wall that spans 246 feet in length and ranges in height from 8 inches to 10 feet. Granite was chosen for the construction of the wall because of its reflective properties; seeing your reflection on the wall is meant to make you feel more connected to the inscribed names of roughly 57,000 Americans who died in the war. The V-shaped memorial aligns with the humility of the widely protested and criticized war in Vietnam. Mitchell writes, "The VVM is anti-heroic, anti-monumental, a V-shaped gash or scar, a trace of violence suffered…" and that it "achieves the universality of the public monument, not by rising above its surroundings to transcend the political, but by going beneath the political to the shared sense of a wound that will never heal, or (more optimistically) a scar that will never fade."[39] The simplistic design of the VVM was criticized originally for being too nihilistic, as it stands in stark contrast to other war memorials, like the United States Marine Corps War Memorial (more commonly recognized as the Iwo Jima monument that features soldiers heroically planting an American flag). Strong opposition to the memorial (completed in 1982) led to a forced compromise two years later, which resulted in the addition of the *Three Servicemen*, a bronze statue featuring three men located near the memorial wall. Nine years after that, yet another installation – the Vietnam Women's Memorial – was added just a short distance away. Today, the VVM is the most visited memorial in Washington, D.C., hosting more than 5 million visitors per year. Statues and memorials are meant to *memorialize*, but values and morals are always changing in society. The evolution of the VVM and its multiple attempts to appease its critics exemplifies how public art is susceptible to political, social, and private interpretations. Let's take another example: Confederate statues.

Since 2015, there have been more than 160 Confederate monuments removed from public spaces. Some of these have been willingly de-commissioned by local governments, but in some cases, protestors have illegally torn down statues and monuments that glorify the Confederacy, which promoted slavery and fought for secession. According to Mitchell's essay, "The fictional ideal of the classic public sphere is that it includes everyone; the fact is that it can only be constituted by the rigorous exclusion of certain groups – slaves, children, foreigners, those without property, and (most conspicuously) women."[40] In other words, a memorial that includes everyone is not possible. But what do we do when memorials, monuments, and statues glorify *unethical behavior*, such as in the case of Confederate statues and monuments that galvanize seditionists and traitors? Perhaps we can turn to The Ethics Centre, which is a nonprofit organization dedicated to bringing ethics to the forefront of personal and professional lives, that has established a set of six criteria that should be used to evaluate whether a public figure should be recognized in the public sphere. These are:

1. They must not be associated with any gateway acts
2. Their achievements must be exceptionally noteworthy
3. Their work must have served the public good
4. The means by which they achieved their work must be ethical
5. They must be the principal driver of the outcomes associated with their deeds
6. The monument must contribute positively to the public commons

These criteria set the bar pretty high for memorializing someone in the form of a statue or monument in the public sphere. Given that no one is perfect, and values are relevant, it may be that we should not be remembering people in this way. Figures that have long been considered unimpeachable, such as Martin Luther King Jr., Gandhi, and Abraham Lincoln, are being re-evaluated through a twenty-first-century lens of values, morals, and ethics. In 2021, the School Names Advisory Committee in San Francisco recommended stripping more than one-third of its public elementary and high schools of their namesakes, which included George Washington, Thomas Jefferson, and Paul Revere.[41] If we look hard enough, we might just find that all humans are flawed, but rather than scrutinize their imperfections, we should judge them on the totality of their actions.

Chapter 2

Evaluating ethics

There are no secrets about the world of nature.
There are secrets about the thoughts and intentions of men.[1]
J. Robert Oppenheimer

Ethical design has existed for as long as humans have made things for a specific use. Ancient tools shaped by early man espoused their own design ethics, but this criticism likely went unmentioned and took a backseat to more pressing matters like safety and survival. Not until recently did modern civilization afford us the luxury of being able to evaluate our technologies not only for their purpose but also for their ethical DNA. In fact, the study of design ethics really did not take off until the late twentieth century; Victor Papanek was one of the first public figures to write about the social effects of designs in his book, *Design for the Real World: Human Ecology and Social Change,* published in 1971. Langdon Winner wrote about "The Politics of Artifacts" in his 1986 book, *The Whale and the Reactor,* and Bruno Latour philosophized about the "scripting" of objects in *Pandora's Hope* (1999). Of course, all of these works were informed by the work of others, including Martin Heidegger, Marshall McLuhan, and Immanuel Kant, to name just a few. And in the modern digital world, when people discuss design ethics, they are mostly referring to the interfaces of computing technologies.

Pop-up ads, "dark patterns," and addictive algorithms raise questions about the intentionality behind the design of our digital interfaces. Often, those intentions are motivated by financial incentives. The phrase "attention economy" refers to how apps and websites profit from user attention and engagement. Just as with currency, there is a finite amount of human attention that we exchange with one another, which makes attention a scarce resource that is highly valued and sought after. Netflix entices you to remain on their platform, not Hulu's. The Instagram app compels you to scroll its feed, not Snapchat's. Subsequently, cognitive tactics are employed to keep you engaged in a desired behavior. This is not a new phenomenon. Supermarkets and clothing stores strategically place popular necessities like milk in the back of their physical spaces, forcing the shopper to bypass other goods on their way and hoping for an impulse buy or two. The user (or shopper) feels like they are in control, but really, the design is stacked against them. In these instances, we can argue that a design that capitalizes on an unwitting subject is unethical. As you will see, *transparency* is a key component of ethical design.

But sometimes, there are just designs that are unintentionally bad, and this does not make them unethical, necessarily. Certainly, we all have our own grievances with poorly designed interfaces, non-working products that were falsely advertised, or just bad craftsmanship. We have all experienced "user error" and even felt frustrated or ashamed because of it. I'm reminded of the Netflix show, *I Think You Should Leave*, which features a sketch where a man is being interviewed for a job. Two men are sitting at a table in a small café, and once the meeting is concluded, one man stands up, shakes hands with the other man, and starts to leave. When he pulls on the door to exit, it doesn't budge, and the man at the table says, "It looks like you have to push." Embarrassed, the other man says, "No, it does both" and struggles mightily to pull the door towards him. The hinges eventually fail, and he breaks the door jamb as he exits the café and says, "See, it does both." The absurd clip is humorous for two reasons. First, it resonates with the viewer; we all have experienced a time when we pushed a door instead of pulling it, or pulled when we needed to push, or sometimes, even stood in front of it waiting for it to open automatically when it was, in fact, a manual door. And second, we all have experienced feeling stupid, embarrassed, or even ashamed when we misunderstand or misuse an object or technology. Some might say this is a user error; the man should've known that the door opened outward, not inward. But this is a *design error*, not a user error. In design terminology, this poorly marked door would be known as a "Norman door," which are doors that fail to clearly communicate to the user whether to push or to pull.[2]

But if we dig a little deeper, we might find that the decision to install a door that opens inward is actually a well-thought-out, intentional design. For one thing, it was an exterior door, and most builders require that exterior doors open inward – both because they do not want to impede foot traffic on the pedestrian walkway in front of the building but also because an inward swinging door has hinges on the interior-facing side of the door, making it more secure from outside intruders. So, there was a good-natured rationale for installing a café door that swings inward. Its lack of signaling to the user makes it a bad design but not an unethical one.

Inversely, an unethical design doesn't necessarily need to be poorly designed. At the time of this writing, Christopher Nolan's *Oppenheimer* is setting box office records, having earned nearly $1 billion worldwide. Naturally, then, the conversation surrounding the use of nuclear weapons has been reignited in recent months. Was the United States ethically justified when the Truman administration dropped two nuclear bombs on Hiroshima and Nagasaki in 1945? Did the use of nuclear weapons convince the Japanese to surrender, thereby preserving many more thousands of lives, or was it an egregious violation of human rights and mass murder of Japanese civilians? These are not questions that I am equipped to answer, nor will I try to hear. I am less focused on the mortal outcomes of the Manhattan Project and the presidential directive to use the atomic bomb and more interested in the ethical decision-making that did (or did not) take place prior to its development and subsequent implementation.

In 1938, three German chemists split the uranium atom. The lead scientist, Otto Hahn, was awarded the Nobel Prize in Chemistry for discovering this process, which would later become known as *nuclear fission* – the catalyst needed for a nuclear explosion. Sensing the danger that would be a Nazi Germany with the capability to produce a nuclear weapon, Albert Einstein signed a letter urging U.S. President Franklin D. Roosevelt to invest resources into the development of the world's first atomic bomb.[3] As history would play out, the Germans failed to overcome the many technical barriers to nuclear fission, and in hindsight, Einstein lamented, "had I known that the Germans would not succeed in developing an atomic bomb, I would have done nothing."[4] He later told a biographer, "I made one great mistake in my life – when I signed the letter to President Roosevelt recommending that atomic bombs be made."[5] Einstein reflected deeply on the ethical consequences that would follow the scientific discovery of nuclear fission and his recommendation to bring forth nuclear weaponry into existence.

There were others who expressed reservations about this technological development. The Committee on Political and Social Problems was formed to address the ethical and moral repercussions of using such a powerful nuclear weapon as an act of war. The final report (known as the Franck Report) was submitted in June 1945 and summed up the committee's position in the following way: "The development of nuclear power not only constitutes an important addition to the technological and military power of the United States, but also creates grave political and economic problems for the future of this country."[6] Nevertheless, in July of 1945, a team of physicists detonated the world's first atomic bomb in the desert of New Mexico. The "Trinity" test, as it would become known, was overseen by J. Robert Oppenheimer, who, at the moment of a nuclear explosion, quoted a line from the Hindu scripture: "Now I am become Death, the destroyer of worlds."[7] He had forever brought forth nuclear weaponry into human existence, and the ethical implications weighed heavily on his conscience for the remainder of his life, as evidenced by his adamant opposition to the development of the even-more-powerful hydrogen bomb. His 1967 obituary in *The New York Times* attributed the following quote to him: "The physicists have known sin; and this is a knowledge which they cannot lose."[8]

The debate over whether or not to develop and then use the atomic bomb plagued even the greatest minds. Whether or not it was morally justified, The Manhattan Project successfully executed its mission and created a weapon of mass destruction, demonstrating that something can be simultaneously well-designed and unethical. To this extent, the matrix below shows the four possible combinations of how objects can be categorized according to their ethics and quality of design.

Table 2.1. A matrix that shows the possible combinations of ethics and design

	Well Designed	**Poorly Designed**
Ethical	A	B
Unethical	C	D

For instance, Norman doors (ambiguously marked doors) might fall under category "B" since they can be ethical but poorly designed. Inversely, the atomic bomb might fall under category "C" because it was extremely well-designed but highly unethical. Designers should strive for category "A," where things achieve the delicate balance of being well-designed and ethical. In fact, it is no longer sufficient to make things simply because we *can*; we must also consider whether we *ought to* develop these things and the impacts that they will have on others to come.

Ethical theories

In order to design ethically, we must first understand what it means to act ethically. So, to accommodate the wide range of interpretations of what ethics is, we might adopt this simple, yet generalizable definition of ethics: *Ethics is mindful wellbeing.* This concise definition rests on two pillars of thought: (1) that we are aware, or *mindful,* of our actions and their intended and unintended consequences, and (2) that *well-being* refers to that of yourself and of others in both the present time and in the future. Within the context of ethical design, ethics refers specifically to the designer's ethical choices (or lack thereof) that have been baked into the design of a product, environment, process, or system and how they impact others. But, providing a theoretical definition of ethics is fairly straightforward. Applying this definition in order to evaluate ethical behaviors, however, is tricky business, as motivations can be interpreted through various lenses. The Markkula Center for Applied Ethics at Santa Clara University identifies six different lenses that may be used to evaluate any particular behavior, and they are as follows: rights, justice, utilitarianism, common good, virtue, and care ethics.[9] The figure below summarizes each lens.

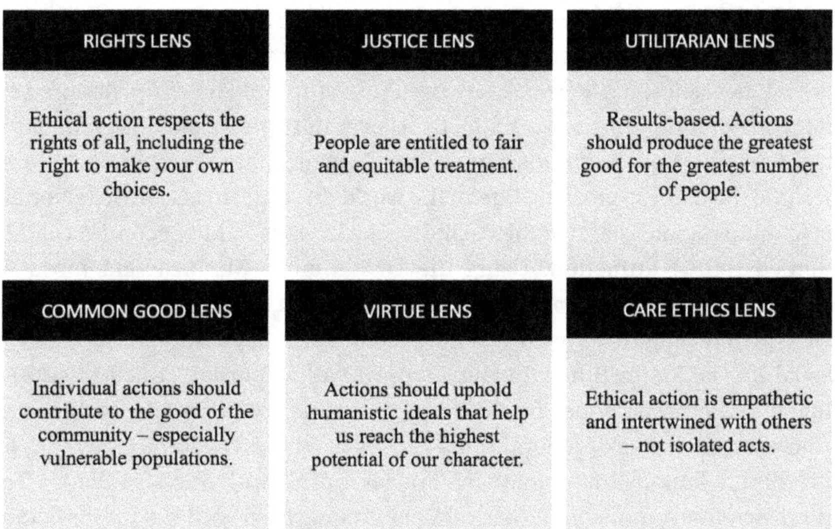

Figure 2.1. A definitional comparison of ethics and morals.

Not all lenses can be applied to all scenarios, and sometimes, these lenses can appear contradictory. Let's put these lenses to the test. As I mentioned in the previous chapter, the federal government mandated that all U.S.-based auto manufacturers include seatbelts in their vehicle designs in the early 1980's. The rationale for this legislation was to save lives; by its nature, this decision was ethical when viewed through the *utilitarian, care,* and *common good* lenses. It might even be considered *virtuous.* And because the law did not discriminate, it would be *just.* However, if we consider the seatbelt legislation through the lens of individual rights, we run into a problem. Is it the *right* of an individual to make his or her own choice to wear a seatbelt, or is it the ethical responsibility of the auto manufacturers and federal government to enforce such laws in order to save lives? This is where the evaluation of ethics enters into a stalemate, depending on your perspective. If you are pro-government, then it is ethical to regulate auto manufacturers if it means saving lives. If you are pro-individual, then this is an unethical breach of privacy and personal freedoms. Ethics is a universal code of conduct and character, and it is not relativistic like morals and values, but ethics can be flexible; it bends towards *what ought to be,* even if that means ignoring less powerful ethical arguments.

If all of this is starting to give you a headache, let's zoom out for a moment. All six of these lenses might be grouped into one of two larger ethical frameworks: deontology and utilitarianism. (There is a third framework, virtue ethics, that I will address later). In order to illustrate the differences between these two main ethical theoretical frameworks, let's use the example of Steven Spielberg's 1993 blockbuster film, *Jurassic Park.* Since it is over thirty years old now, it might be useful to give a quick synopsis of the movie, which, at the time of its release, became the highest-grossing film in the world. In short, an eccentric billionaire develops a remote island in the Pacific Ocean in what was to become a wildlife-themed tourist attraction. Using the DNA from a Mesozoic-era mosquito preserved in amber, a team of genetic scientists is able to re-animate dinosaurs using genetic engineering and cloning techniques. In addition to its stunning visual effects (for the time), the film grapples with important ethical dilemmas, namely, whether science should override natural selection and evolutionary processes, such as extinction. In the movie, actor Jeff Goldblum plays Dr. Ian Malcolm, a famous mathematician known for his work in chaos theory. This interdisciplinary, mostly theoretical field is concerned with the unpredictable course of events. A popular example used within chaos theory is the concept of the "butterfly effect," which suggests that a small change in an initial condition can cause a chain reaction, resulting in a larger (often undesirable) outcome. Dr. Malcolm gives a foreshadowing speech at the beginning of the film, where he questions the ethical consequences of re-animating extinct dinosaurs for a theme park:

> Your scientists were so preoccupied with whether or not they could that they didn't stop to think if they should. Science can create pesticides, but it can't tell us not to use them. Science can make a nuclear reactor, but it can't tell us not to build it![10]

Malcolm is very clearly delineating between what we *can do* and what we *ought to do.* And in doing so, he is expressing the two pillars of ethics, as defined earlier: (1) he is raising *mindful* awareness and signaling concern over the ethics of genetic engineering, and (2) he is suggesting there may be future consequences that affect the *wellbeing* of others. Without saying so explicitly, Malcolm is a utilitarian – also known as a consequentialist – because he is concerned about the repercussions of de-extinction. The main ethical frameworks of deontology and utilitarianism can be thought of as opposite ends of the ethics spectrum.

Table 2.2. The two main ethical frameworks: deontology and utilitarianism.

Main Ethical Frameworks	
Deontology	**Utilitarianism**
Intent-oriented	Results-oriented
Focuses on the *act* itself.	Focused on the *consequences* of the act.
Deontology (-*deon,* Greek for duty) is duty-based. For instance, the Ten Commandments outline a set of consistent practices that should be dutifully followed, regardless of its consequences.	Utilitarianism, or *Consequentialism,* aims for the collective good. However, this approach can overlook justice or compassion for those marginalized (e.g., South African Apartheid).
- EXAMPLE -	
Killing is always wrong.	Killing an active mass shooter to prevent the loss of life is acceptable.
- MAXIM -	
Let justice be done, though the heavens may fall.	The greatest good for the greatest number.
- NOTABLE FIGURES -	
Immanuel Kant, René Descartes	Jeremy Bentham, John Stuart Mill

There are many different theoretical frameworks for ethics; the two main ones, deontology and utilitarianism, illustrate a dichotomy of sorts. To better understand the difference between the two, let's apply them to a real-world example such as healthcare. The area of medical ethics constantly faces moral and ethical dilemmas that have literal life-and-death consequences. In this medical ethics context, a deontological physician focuses narrowly on treating the patient; a utilitarian physician treats the patient but also considers the impact that this treatment has on the rest of society.

The avatar for deontological ethics is Immanuel Kant, an eighteenth-century German philosopher. Kant is credited as the father of deontological ethics, which is sometimes referred to as Kantian deontology. His strong religious faith impressed upon him the value of each and every human life. Therefore, deontological ethics "emphasizes the value of being a human being, underlining the principles of respect for autonomy, beneficence, non-maleficence, and justice,"[11] even when this means putting yourself in danger. During the COVID-19 pandemic, for instance, healthcare professionals endangered their own welfare by treating those who were afflicted with the unknown and misunderstood coronavirus. This exemplifies how moral *duty* and *obligation* inform the deontological ethics ideology. Even in the face of a global pandemic, the Kantian deontologist will try to preserve the well-being of every individual, no matter the cost.

The contrarian view is rooted in utilitarian ethics, which is attributed to the eighteenth-century English philosopher, Jeremy Bentham. His major work, *An Introduction to the Principles of Morals and Legislation*, laid out his view of utility, arguing that ethical decision-making is fundamentally a choice between an individual's pleasure and pain rather than a moral obligation or duty to certain inalienable rights. Bentham emphasized the "maximization of happiness"[12] and translated this to mean that an ethical action produces the greatest good for the greatest number of people. No matter the circumstances, the correct action is the one that benefits the most, even if this disregards human autonomy and welfare. Using the example of the medical ethics from above, a utilitarian approach to the COVID-19 pandemic would have been to identify the area of the outbreak and to contain it by any means necessary – including total quarantine – to save the rest of society from becoming infected. This is precisely the plot of the 1995 film *Outbreak*, in which a deadly virus infects several residents in a small California town. Government officials decide to completely eradicate the town and its residents by dropping a bomb in order to prevent transmission of the virus. Bentham would have cosigned this decision, reminding us that "as long as there are these good effects which are, on balance, better than the effects of any alternative course of action, then the action is the right one."[13] Clearly, deontological and utilitarian ethics are at odds with one another, and both approaches have their shortcomings.

We might also contemplate these ideologies through a design ethics perspective. Would you agree that we should embrace renewable energy sources as an alternative to non-renewable fossil fuels? If so, you might already know that solar wind farms can provide a powerful energy source at a reasonable cost. In fact, the United States Geological Survey (USGS) reports that across the span of one month, a single windmill turbine generates enough electricity to power roughly 940 homes.[14] Nevertheless, a Kantian deontologist approach would oppose such infrastructure, citing that the imposing windmills kill thousands of birds and the solar farms used to store the energy disrupt animal habitats. Conversely, according to utilitarian logic, we should erect as many windmills as are needed to power the entire country, regardless of whether these massive structures visually contaminate our landscapes, create noise pollution, or even infringe on cultural heritage sites and national parks. As you can see, the problem with utilitarianism is that if we always focus on maximizing happiness for the greatest number of people, this could incentivize unreasonable behavior.

Ethical theories also might be thought of as either absolute (universal) or situation-based. As these names suggest, an absolutist approach to applying an ethical theory "assume[s] that a particular set of ethical principles is always valid, regardless of the surrounding circumstances,"[15] whereas a situation-based approach considers changing variables and prioritizes ethical theories accordingly. The most reasonable approach to evaluating any situation and/or behavior might instead be a combination of the two. Aristotle first posited that ethical virtue, or *hexis*, depends on both the situation and moderation. For instance, the New York City-based Major League Soccer (MLS) team, NYCFC, currently plays their matches in Yankee Stadium, an outdoor arena that has a seating capacity of more than double the team's average attendance in 2021.[16] In 2027, New York City's first soccer-specific stadium will be completed at an estimated cost of $780 million and will be able to accommodate 25,000 fans. To Aristotle's point, it would be unethical (and unjustifiable) to build a stadium that holds 100,000 people, considering the sizable building footprint that would be required in one of the most expensive real estate markets on the planet. Likewise, it would be inadvisable to build a smaller, cheaper stadium that only accommodates 5,000 fans, as this would likely create a premium on ticket prices and exclude much of the fan base from watching their team play. In this example, it is not sufficient to think solely about the ethical design of the stadium; it also requires an awareness of the situational variables and how moderation can be achieved. This is sometimes referred to as the "Goldilocks" zone: not too much and not too little. And sometimes, it's the players – not the stadium – that create unpredictability; Inter Miami recorded the lowest average attendance of all MLS teams in 2021, and then they signed Lionel Messi, who is largely considered the world's best player. Ticket prices to attend an Inter Miami game in the DRV Pink Stadium, which only holds about 19,000, surged more than 1000%, and seats at midfield skyrocketed to $23k on the secondary market to watch Messi's first game in Miami.[17]

Deontology and Utilitarianism are *ethics of conduct*. This is separate from a third area of theory: Virtue Ethics, or the *ethics of character*. Whereas the ethics of conduct responds to the question, "What ought I do?," the ethics of character responds to the question, "What's the best person to be in this situation?" You can see the difference here using the example from above. Suppose you happened to be a season ticket holder for Inter Miami before they acquired Lionel Messi. Your $30 ticket just increased its value to nearly $600, and you find that you are unable to attend Messi's debut game due to a scheduling conflict. A deontologist or utilitarian, recognizing that the situation has changed, might resell this ticket and profit handsomely because it maximizes happiness and results. A virtuous person, however, might rebuke this as a blatantly greedy cash-grab, and instead, give the ticket to someone less fortunate. Although being virtuous is a desirable character trait, it is not necessary for creating an ethical design. Rather, moderation is preferred. In the earlier example of the divided bench, an ethical design simply promotes inclusivity and rejects the hostility baked into the design that prevents the unhoused from sleeping on benches. A virtuous design, however, might suggest that these benches go even further and include sheets and a pillow to better accommodate the unhoused. Virtue ethics says more about the designer than the design. And since this is a book on ethical design and not virtuous behavior, we will focus mainly on deontology and utilitarianism as the two main ethical frameworks.

So, this begs the question: Would you describe yourself as deontological or utilitarian? Are you more concerned with *intent* or with *results*?

The trolley problem

There is a famous thought experiment that has been circulating among moral philosophers since Philipa Foot first wrote about it in her 1967 paper, "The

Problem of Abortion and the Doctrine of Double Effect," which appeared in *The Oxford Review.* The "trolley problem," as it would become known, has many variations but essentially poses the central question: "When does an impermissible act become permissible?" Of course, the thrust of this paper was oriented toward the doctrine of double effect, specifically as it pertains to abortion rights. But it is Foot's brief description of the trolley problem that continues to perplex philosophers to this day. Indeed, the trolley problem persists even for designers who are faced with these same ethical and moral dilemmas as they develop modern technologies like autonomous (self-driving) vehicles, artificial intelligence, and biotechnology. The hypothetical scenario goes like this:

> There is a runaway tram, or trolley, barreling down the railway without the ability to brake. Unfortunately, on the tracks ahead, there are five workers – all of whom will be killed if the trolley continues on its course. However, the trolley driver can choose to throw a lever that would redirect the trolley to a secondary track (known as a "side spur"), and by doing so, this would divert the trolley away from the five workers on the main track but kills one man standing on the side spur. As the trolley driver, do you throw the lever to divert the trolley to the side spur, killing one man, or do you let the trolley continue straight on the tracks, killing five men?

If your intuition is to throw the lever and divert the trolley towards the side spur, killing one person instead of five, then you would be in agreement with the majority of respondents. If the ethical solution was simply to mitigate the loss of life, then this would be an easy answer. But ethical dilemmas are messy. Foot goes on to illustrate this further by giving another, more elaborate hypothetical scenario:

> There is a judge who is being pressured by an angry mob to find someone guilty of a heinous crime that has been committed in the community. Unfortunately, the true culprit of the crime is unknown. If the judge does not sentence a guilty party, the bloodthirsty mob has vowed to take the lives of their five hostages. Does the judge frame an innocent person as the patsy for the crime and sentence him to death, thereby saving the lives of the five hostages?

In this scenario, if you responded "no" – the judge should not wrongfully sentence an innocent man to death, even if it means rescuing the lives of the five hostages – then you would be in agreement with most people.[18]

Supposing your decisions aligned with the majority of others in both scenarios, you would have diverted the tram in the trolley problem to kill one instead of five men, but in the judge problem, you would have refused to kill one man to save five others. These thought experiments show that there are extenuating circumstances that must be weighed in every ethical decision. For instance, a strict utilitarian would abide by the maxim that we should take any action necessary to reduce the loss of life; in other words, always choose the scenario where the fewest number of lives is lost. But the scenarios above show us that it is not quite that simple.

Consider a third variation of the trolley problem in which a runaway tram will kill five men tied to the tracks unless it is stopped. Crossing above the tracks is a pedestrian footbridge, where you happen to be standing with a "fat man."[19] You are now faced with a decision: Do nothing, and the trolley will run over the five men, killing them. Or you could push the fat man from the footbridge and onto the tracks, which would kill him, but it also would stop the tram and save the five men. If you are like most respondents, your answer flips from the original trolley scenario when it felt permissible to kill one man on the side spur if it meant saving five lives. But in the fat man scenario, it feels ethically impermissible to push one man off a footbridge to save five others. Even though the outcome is the same – five men are saved – the means by which the men are saved determines the permissibility of your actions.

Judith Jarvis Thomson would publish her own response to Foot's moral test in her 1976 paper, "Killing, Letting Die, and the Trolley Problem." In it, she argues for the moral distinction between *killing someone* and *letting someone die.* Thomson would subsequently tweak the trolley problem to make the central figure a trolley passenger (and later, a bystander) who did not set the trolley in motion, nor was she responsible for driving it. Would the bystander be responsible for throwing the lever and diverting the trolley to kill one instead of five, or should she do nothing and remain an innocent bystander (watching idly as the five are killed)? Thomson would argue that "negative duties do not trump positive duties"[20] and that throwing the lever would be considered impermissible for a bystander to perform. If you despise hypothetical scenarios like these, then you might be inclined to simply do nothing. After all, you didn't ask to be put into this position to save someone's life. But neglecting to make a choice is, by default, a choice. Not throwing the lever on the trolley or not throwing the man off the footbridge does result in a consequence that you could have changed.

From a legal perspective, there are liability and negligence laws that stipulate the degree to which a passerby has the duty or obligation to intervene in a situation where another person is in peril. This is known as the "duty to rescue," and its requirements vary by state and by country. In the United States, only a few states mandate that citizens render aid, or at the very least, alert authorities when someone requires medical attention. For example, in 2017, five teens in Florida witnessed a man who was drowning in a lake; instead of jumping in and rescuing him, they narrated his death as one of the boys recorded the event on his smartphone and then later posted it to social media.[21] Because Florida does not enforce a duty to rescue law, the teens were not found liable for the man's death, but the preventable tragedy reinvigorated debate around the moral duty to provide help to others and the legal culpability if there is a failure to do so. In the United States, the common law no-duty rule stipulates that a person does not have a responsibility to intervene in a perilous situation.[22] By contrast, in France, the legal expectation for the duty to rescue is exceptionally high. Perhaps the most famous example of a criminally negligent nonintervention was the death of Princess Diana in 1997. The Mercedes Benz that Diana was traveling in was speeding at nearly double the posted speed limit on a roadway in Paris, France trying to outrun the paparazzi when it collided head-on with a stone pillar that supported the overpass above. Sadly, three of the four passengers, including Diana, Princess of Wales, would later succumb to their injuries. The accident ultimately was ruled to be the fault of the driver, though the nine photographers who arrived on the scene and took photos rather than render aid would be criminally charged with manslaughter, due to the strict duty to rescue laws in France.[23]

When it comes to evaluating ethical *actions*, most of us can differentiate between right and wrong, just and unjust, and humane and inhumane. For instance, most would agree that it was wrong for the teenage boys in Florida to watch another person drown instead of help. Similarly, it was inhumane of the paparazzi to photograph the gravely injured Princess rather than to seek help. But ultimately, there was no legal consequence for these actions; the teenage boys in Florida and the paparazzi in France beat their criminal charges (though the nine photographers would be found guilty of invasion of privacy). Both of these examples are extreme cases of negligence, but I use them to illustrate an important point: you are not *legally required* to be a decent person who makes ethical decisions. In fact, Thomas Hobbes (whom I discussed in Ch.1) argued that – in the absence of regulation – people resort to a competitive "state of nature" in which they look out only for their best interest and not for others.[24] It is your personal freedom to be unethical, but that doesn't mean that you ought to be.

Just as we can evaluate our actions ethically, so too can we evaluate design. In his essay, "Toward the Spiritual in Design," Victor Papanek describes the ethical dilemmas that all designers must face. Whether it be internet technologies, a stylish lamp, a self-driving car, or a solar farm, everything has a "spiritual value" deeply embedded within that consists of the "*intent of the designer* as well as the *intended use* of the designed object [emphasis added]."[25] Spirituality, in this sense, refers to positivity in design. According to Papanek, an object designed with the explicit intent to harm – for instance, a fighter jet – "could not have beauty – let alone spiritual value." We might extend an object's spirituality to its ethical posture; a spiritually positive design is also an ethical design. To evaluate this spirituality in design, Papanek proposed a set of six questions:

- Will the design significantly aid the sustainability of the environment?
- Can it make life easier for some group that has been marginalized by society?
- Can it ease the pain?
- Will it help those who are poor, disenfranchised, or suffering?
- Will it save energy or – better still – help to gain renewable energies?
- Can it save irreplaceable resources?

Ideally, an ethical design would answer "yes" to each of these evaluative questions, but in reality, a technology or product design might only satisfy one of these criteria. This is what makes evaluating ethical designs so difficult. A new technology might be developed to ease pain, but in doing so, exploit a non-renewable energy source or negatively impact a marginalized group of people. Indeed, problem-solving depends heavily on the perspective of the individual.

Often, design responsibility falls onto the engineer of the product, who may or may not have had ethics training or who may or may not be interested in developing an ethical product. In fact, the design of a product highly reflects the image of the designer. An article in *Philosophy & Technology* describes how designers and engineers with different backgrounds might approach problems differently, according to their own ethics, morals, and values. For example, consider a scenario where a designer is asked to build something to keep birds off a piece of property. Some solutions might look like this:

- A designer with a background in construction might suggest using spikes on rims and poles where birds like to sit.
- Someone with a background as a falconer might suggest getting a falcon to scare off the other birds.
- A sound engineer might suggest using high-pitched noises to keep birds away.
- An environmental activist might suggest catching and relocating the birds to a safer location.[26]

There is no "right" answer here, but some of the above options present more ethical designs than others. Arguably, the environmental activist has the most humane approach, though this also might be the most expensive solution. Installing spikes on all flat surfaces or high-pitched noise machines to deter birds might be a thrifty approaches, but they create a hostile design. Importantly, we must consider all variables when creating, proposing, and adopting technological solutions and designs to promote spiritual positivity in our built environment. As Philipa Foot's trolley problem has shown us, every decision has its consequences – even the decision not to decide. And the design of a technological product, environment, process, or system is a series of collective decisions made by groups of individuals. So, then, it is imperative that we are not only mindful of the ethical posture of the things that surround us but also of the things that are yet to come.

Critical design

When you picture the future, what does it look like? Is it filled with flying cars, hoverboards, self-drying clothing and self-lacing shoes, virtual reality eyewear, and dog-walking robots? This was precisely how the fictional year 2015 was represented in the 1989 film *Back to the Future Part II,* directed by Robert Zemeckis. Although the movie is set nearly thirty years into the dystopian future, Zemeckis accurately visualized today's culture to be highly socio-technological with the ubiquity of video conferencing, virtual gaming, bionic implants, and was only one year off from predicting that the Chicago Cubs would win the World Series for the first time since 1908.[27] Many of Zemeckis's predictions remain absurd and disconnected from reality, but artists, architects, writers, and futurists have long imagined future scenarios that are rife with problems and possibilities, most often stemming from an inevitable deep embodiment of technology into the human condition. Artifacts such as dystopian novels and films serve to challenge the status quo and can provoke important ethical questions through practical and conceptual design.

The difficulty with writing or directing a dystopian piece of fiction is that it requires some level of prediction-making, some of which may come true and most of which never will. We are nowhere near Zemeckis's futuristic vision with flying cars and hoverboards. But sometimes, films can be eerily prescient. Spike Jonze's film *Her* (2013) imagined a future where our smartphones are artificially intelligent and ingrained into nearly every aspect of our lives. The main character, Theodore Twombly, struggles with human relationships and instead prefers the company of his AI-powered futuristic smartphone, Samantha, with whom he eventually falls in love. The film grapples with how technology will become a proxy for human-human interaction before it ultimately culminates into what can best be described as the *technological singularity*: a hypothetical point in time in which technology becomes autonomous and recursive, no longer requiring human input or programming to expand. In the ten years since its release, *Her* has gotten a lot of things right, including the miniaturization of wireless earbuds, the advancements in voice-recognition software, and the sophistication of AI. But the relevance here is not the accuracy of the technological realism that is presented in the film, but the provocation of what might happen if these things did exist. Zemeckis was imagining how our futures would be made more convenient by technology; Jonze was imagining how technology would replace our humanity.

In both examples, the directors had to conceptualize technologies and whole environments in which their stories could take place. In doing so, they were engaged in a practice of *critical design* – a term first introduced by Anthony Dunne in his 1999 book, *Hertzian Tales*, meant to "challenge narrow assumptions, preconceptions and givens about the role products play in everyday life. It is more of an attitude than anything else, a position rather than a method."[28] A major criticism of critical design is that it is viewed as being skeptical and cynical of mass-produced things in popular culture, taking a pessimistic and sometimes nightmarish view of technology. But in an online Q&A forum with critical designers Anthony Dunne and Fiona Raby, they dispelled the accusation that critical design is strictly negative, saying: "Is critical design a bit dark? Yes, but not for the sake of it. Dark, complex emotions are ignored in design, most areas of culture accept people are complex, contradictory, even neurotic, not design, we view people as obedient, predictable users and consumers."[29] Dunne goes on to levy further accusations at modern design, adding that it is "not engaging with the social, cultural, and ethical implications of the technologies it makes so sexy and consumable."[30] The critical design approach to modern problems can be a valuable exercise in skepticism, but it is not just about shutting down creative ideas, technologies, and products; critical design is meant to question the unquestioned and to ignite discussion and reasoned debate about the ethical nature of design. Whereas Zemeckis had an optimistically favorable view of how technology would make our lives easier and more convenient, Jonze saw technology pessimistically as a threat to the human condition. Zemeckis was hopeful; Jonze was dark.

Critical design has spawned other terms like *design fiction* and *speculative design*, which are more prescriptive than descriptive. Critical and Speculative Design, sometimes abbreviated as CSD, is often referred to as a field, a discipline, or method, when in reality, it is more of a philosophy.[31] Essentially, CSD is a deliberate process that imagines future scenarios in which a technological product, environment, process, or system might be adopted and the ripple effects that it might then cause. A novelist or filmmaker who is composing a futuristic story is engaged in critical design, and so too are designers, architects, and engineers who are creating the products and environments of the future. According to an article on UXPlanet.org, "A typical design process goes like this: idea generation, research, ideation, iteration, and testing. But with speculative design, we add two extra steps: exploration and visualization."[32] Instead of stopping at the question, "How might this work?" designers might use speculative design to extend the question to: "How might this work for ***x***?" – where ***x*** stands for different groups of users. To further illustrate this point, here is a quick example: As part of a climate-friendly pledge, energy-saving lights were installed in the gymnasium restrooms at my kids' school a couple of years ago. The lights remain off until a motion sensor detects when someone has entered the restroom. Once motion is detected, the lights turn on. But after one minute, if no motion has been detected in the room, the light automatically shuts off. My youngest son, who is seven years old and stands about 45" tall, has gotten trapped in the pitch-black restroom on multiple occasions because the motion detector was installed at the building code height of 48" on the wall. Installing energy-saving lighting in the bathrooms shows a noble effort to conserve energy, but designing and installing them in this way excludes younger (and shorter) populations, which is problematic in a K-12 school environment.

Conducting a stress test that imagines future design problems is sometimes referred to as a "dark reality session" where all possible effects are envisioned.

The interaction design group, A List Apart, has articulated some important questions that should be asked during this speculative design phase of ideation that can help anticipate future ethical issues in design:

- What is the lifespan of your product?
- What if the user base will be in the millions?
- What are the long-term effects on the economy, society, and the environment?
- Who benefits from your design?
- Who loses?
- Who is excluded?
- How could your design be misused?[33]

And critical design pioneer Anthony Dunne adds the question:

- Who else is going to be touched by this innovation?[34]

In critical design, an ethical designer becomes strictly utilitarian, intently focused on the consequences of the design and the subsequent impact that it will have on users, non-users, and the surrounding environment. Most times, however, it only becomes evident that a technology or product is unethical after it has already been adopted. A popular podcast titled *Should This Exist?* Produced a total of 22 episodes from 2019 – 2020 that focused on innovations like bioartificial kidneys, nuclear power, emotional robots, synthetic meats, and killer drones to discuss how technologies often are used in good and in bad ways and to ultimately rule on whether it is advisable to bring this technology into existence. This answer is not always clear, however, because everything has a battery of different uses that obfuscates the ethical direction of design.

All technologies have multiple uses – these are intended and unintended. This flexibility was termed *multistability* by the technology philosopher Don Ihde. A commonly used example is a chair. While it is intended for sitting (it actually *affords* sitting, as we will see in the next chapter), a chair also might be used for standing on to change a lightbulb, for propping open a door, or for throwing through a window to escape a room that is on fire. Another example is the smartphone. This dynamic, multifunctional tool has many cognitive applications; it is a web browser, a camera, a navigational aid, a source of news and entertainment, and much more. But it is a versatile physical object as well. The smartphone can be a paperweight, a remote control for the television, or it can become a weapon if thrown at someone. As postphenomenological research tells us, there are a number of factors that influence how we perceive and use technologies, and this adds up to the *intentionality* of technology, or how we interpret and intend to use something. Often, culture plays an important role in how we situate objects in our environments. In his book, *The Timeless Way of Building*, Christopher Alexander talks about how the sidewalk is a place for congregating, socializing, and playing music in India, but in New York City, the sidewalk is a bustling thoroughfare for speedwalking pedestrians, where one dare not sit or stand still. It is important that we account for all uses and contexts when designing ethically.

Unethical designs are all around us, and often, it is up to the user to identify and react to them. Alan Cooper is a leading figure in interaction design and computer programming, and he now develops user-centered software that emphasizes the human interaction experience with technology. In a 2018 keynote speech at the Interaction 18 conference for designers, Cooper warned that our modern technological innovations are not simple systems. Our technologies and products are connected, making them networked and complex. Therefore, he argued, the problems created by these systems also are distributed and complex. He likens our built environment (which includes all manmade technologies and products) to the hull of the fatally doomed ship, *Titanic*. Unethical designs creep into the built environment just as water enters the hull of the ship, causing it to sink. But unlike the actual *Titanic*, which suffered a single catastrophic gash in its hull, our built environment has a "constellation of tiny ones... a million microscopic laser-drilled holes" that eventually cause the hull to flood. And so, according to Cooper, our focus should be to "prevent these tiny leaks, rather than trying to predict and then stop a single catastrophic event." His view is not that these unethical designs are evil, but that because technology and society evolve at such a rapid pace, well-intentioned design can quickly turn bad, unethical, or even dangerous. Cooper punctuates his talk by reminding the audience that "Our innocence isn't sufficient."[35] We are all consumers of technology, but at what point do we become complicit in unethical designs? When do we become unethical *consumers*?

Unethical products are defined as "those goods and services which any stakeholder believes may damage society as a whole. This may be the result of the firm operating in an inappropriate manner or because the actual product is potentially damaging to it users."[36] It is likely that you use unethical products each and every day; Apple and Nike are just two examples of major brands that have been accused of unethical labor practices. But it is unreasonable to go through life evaluating every object with which we interact. Instead, ethical design is a shared responsibility among designers, companies, employees, the environment, the local community, and the customer. Below is a checklist that I compiled, which shows the various stakeholders (designer, user, society) and the specific criteria that each should strive to meet in order to achieve ethical design. Violating any one of these areas should trigger a further evaluation of ethics.

Table 2.3. Checklist for Ethical Evaluation of Systems, Products, and Processes.

EVALUATIVE QUESTIONS		
Designer	***The design...***	
Accountability	can be traced back to a specific person, people, or institution.	
Transparency	clearly articulates the intended functionality and potential risks of the design.	
Safety	considers the physical, mental, and psychological well-being of the user.	
User	***The design...***	
Value-Sensitivity	is inclusive and welcoming of individuals, bridging wellbeing and social responsibility.	
Privacy	communicates what information about the user can be shared with others.	
Accessibility	is mindful and empathetic towards all users' abilities and limitations.	
Society	***The design...***	
Economic	helps those who are poor, disenfranchised, or suffering, or will it ease economic pain for a marginalized group.	
Environmental	preserves irreplaceable resources, help to gain renewable energies, and not do irreversible damage to the natural environment or compromise future generations.	
Sustainability	contributes to – and does not detract from – the sustainability of the environment and natural, non-renewable resources.	

Admittedly, the standard for ethical design is remarkably high, and the use (or misuse) of technologies cannot always be foretold. Critical design can only account for as much as we are able to anticipate, which sometimes is not enough. But when we learn of unethical products and practices, it is our duty to re-evaluate their use at the very least. Yuval Noah Harari is a bestselling author whose award-winning works include *Sapiens: A Brief History of Humankind*, *Homo Deus: A Brief History of Tomorrow*, and *Nexus: A Brief History of Information Networks from the Stone Age to AI*. A common theme that runs throughout Harari's books is that emerging technologies usually give rise to new power structures that shape social order and force humans to see themselves in new ways, and this raises genuine ethical concerns. Importantly, Harari recognizes that "Technology only creates new opportunities; it is up to us to decide which ones to pursue."[37] You would be hard-pressed to find many products and technologies that satisfy all the criteria in the checklist above, so does that mean that you should never interact with anything that falls short of ethical perfection? No. Some moderation is required. We will never be able to defeat an abstract concept like unethical design, but we can live mindfully and make ethical choices when possible. As Alan Cooper has suggested, we cannot claim "innocence" any longer.

Case study: Location tracking devices

In 2013, a technology company called Tile began crowdfunding its simple invention: a Bluetooth-enabled device, about the size of a poker chip, that could be affixed to any personal belonging and its location tracked using its native app on a smartphone. Tile launched its tracker in 2014 at the affordable price of $25 per unit and has a reported 40 million users.[38] In 2021, the company was acquired for $205 million by Life360™, a family social networking app that specializes in family communication and location tracking.[39] That same year, Apple released the AirTag, and Samsung introduced the Galaxy SmartTag. While each tracker differs in its own way, both aesthetically and functionally, they all serve the same purpose: emit a wireless signal to its user for real-time location tracking of any object. Naturally, you can probably see the ethical complication here.

The miniature trackers are hardly noticeable and can be fixed to personal belongings like keychains, purses, electronics, bikes, vehicles, pets, and anything else that you want to keep track of. Users then can enable location tracking and use a native app to view the whereabouts of each individual tile or tag. But what if the desire to track something (or someone) is more nefarious?

There have been numerous reports of Russian and Ukrainian soldiers using mobile devices as trackers to identify the locations of their enemies, and in some cases, to sabotage humanitarian aid efforts.[40] The multistability of the Bluetooth-enabled tracking device is a serious matter. Yes, the tile can be helpful in tracking down the whereabouts of a lost pet, but it also can be used to track an ex-partner or spouse without their consent. Last year, an article in *The Washington Post* described how these trackers have been found in coat pockets and taped to the undercarriage of cars,[41] and *The Guardian* chronicled stories of celebrity actors and models who have found trackers planted on their person.[42] The latest stalking victimization report from the U.S. Department of Justice has created an entire subcategory – "stalking with technology" – that distinguishes the method by which the stalker offends the victim and includes texting, voicemails, emails, and internet or social media messages. Further subdivided is the number of victims aged 16 or older who were spied on or monitored using technology (610,180) and who were tracked using an electronic device or application (394,000).[43] These figures now outpace the number of offenders who waited in a specific location to encounter the victim. In other words, tile trackers have made it easier than ever to stalk victims using technology.

The rise in the number of unwanted tracking incidents has prompted some updates to these devices. For example, Apple has introduced what it is calling "an industry first"[44] for built-in tracking security. The AirTag now sends a notification to the iPhone app when it identifies an unregistered device that has been traveling with the user. Additionally, no data of physical location history is stored on the device. Similarly, Tile by Life360 ™ and Samsung have taken steps to prevent unwanted tracking. The companies have unveiled its "Scan & Secure" and "Unknown Tag Search" features, respectively, that work as a scanner tool to sweep an area for unwanted tracking devices, though these must be viewed through native apps that require a download (hence, the transfer of personal data to these companies). Yet these privacy features illustrate the ethical dilemma that these tracking devices have created. In order to combat unwanted tracking and to remain safe, each person must be (1) informed and aware that this is a problematic use of the technology that exists regardless of whether they are users of the technology and (2) have the knowledge, ability, and motivation to proactively deter this technology from being used against them in a harmful way. From an ethical design view, this is troublesome because the technology places the burden of safety onto the user to ensure that they are not being stalked, rather than preventing the offender from stalking.

This case study underscores the ethical implications of tracking-enabled devices and whether the convenience of digitally tracking our things is worth ushering in the potential for unwanted voyeurism, or worse. Professor Richard Mason wrote an oft-cited article titled "Four Ethical Issues of the Information Age" that appeared in *Management Information Systems Quarterly*. In it, Mason articulated four ethical issues that pertain to technology: privacy, accuracy, property, and accessibility. Mason warned that invasions of privacy are an affront to human dignity and that they "creep up on us slowly."[45] And, as we continue to contribute our personal data to new technologies, we feed into a rapidly expanding network of connected devices known as the Internet of Things (or IoT). This term first appeared in 1999, and by 2008, there were more connected devices on the planet than there were people.[46] The amassing of smart devices has aggregated our privacy data to form new relationships. According to Mason:

> The bringing together of disparate data makes the development of new information relationships possible. These new relationships may be formed, however, without the affected parties' permission. You or I may have contributed information about ourselves freely to each of the separate databases but that by itself does not amount to giving consent to someone to merge the data, especially if that merger might reveal something about us.[47]

This scenario becomes even more alarming when you consider that tracking devices can easily target and exploit already marginalized and vulnerable populations such as young children, the elderly, members of the LGBTQIA community, and the socioeconomically disadvantaged. As a result, the author pleads for a "new social contract" in which "information systems should not unduly invade a person's privacy."[48] It might surprise you to learn that Mason published this article in 1986 – seven years before the launch of the internet as we know it today.

As we continue to flood our environments with more and more trackable technologies, we must stop and consider the outcomes. A speculative design exercise would most certainly reveal questions about how these tools can be misused and how they can compromise individual privacy. But the multistability of these trackers clouds their ethical design; for every nefarious use of the AirTag, you can cite multiple benefits. This is precisely the issue with evaluating ethics: technologies are good and bad, positive and negative, useful and harmful. But this does not mean that they are *neutral.* Through their designs, location-tracking objects compel us to behave in specific ways, which may discount the well-being of those being tracked.

Chapter 3

The Built Environment

We shape our buildings and afterwards our buildings shape us.[1]
Winston Churchill

Each day, from the moment we wake up to when we lay our head on a pillow to go to sleep at night, we encounter design. These designs can be *restrictive*, such as with a locked door (which prevents us from entering), a flat tire (which prevents us from driving), and a poor wireless signal (which prevents us from communicating). But these designs also can be *affording*, such as with the pull knob on a cabinet door (which affords opening), a paved road (which affords driving), and a well-placed cellphone tower (which affords communication). Many of these designs go unnoticed until something does not work properly or does not allow us to successfully perform a task. In fact, technology philosopher Langdon Winner described this as "technological somnambulism" – where users "sleepwalk"[2] mindlessly through their use of technological artifacts and the deliberately designed built environment that surrounds them. We take for granted the technological advancements that have been adopted by previous generations and then scaffolded onto. Consider this: for most of human history, fire-making was an essential skill for survival; without the knowledge and ability to start and maintain a fire, humans would have found it challenging to survive in colder climates. Today, however, HVAC systems, thermostats, and insulated walls afford us the ability to dial in a precise temperature for comfortability and warmth even in the harshest of climates. I doubt any of us reflect on the marvelous technological evolution that has unfolded in the timeline of human history each time we click our thermostat up or down by one degree. In fact, we probably only ever contemplate the luxury of forced air when our HVAC system stops working. Martin Heidegger used the term "present-to-hand"[3] to describe the brief epiphany that we experience when a technology fails to perform as expected. We constantly anticipate our interactions with technological objects and with our surrounding environments, and only when they fail do we awake from our slumber. Jesper Aagaard builds on Winner's idea of somnambulism by referring to humans as "passive subjects" when he states the obvious, yet critically important point that: "Not all our technology use is governed by conscious control."[4] This does not mean that we are controlled or even manipulated by our technologies, as a technological determinist might argue. Instead, we might take the softer perspective that we are nudged or even compelled to certain actions because of the combination of deliberate, intentional design and our generally oblivious and unmindful behaviors. To be clear, we are not victims of our surroundings, but we often are held captive to their designs and to their ethical makeup. George Nelson's 1977 text, *How to See*, is a seminal piece of writing on how design is both pervasive and influential. Nelson writes, "Every object in any environment may be considered a nonverbal message input, subject to decoding by any suitable receiver within range."[5] He muses on the message design of objects like street signs, public art, and letterforms, and of larger societal designs that include infrastructure and urban design. But his central point is that we must learn to interpret our environment through seeing design. Nelson argues that "if an ability to read words helps us get around, make a living, and add interest to our lives, so does seeing." On the contrary, we are often passive participants within the built environment. Take a look at your surroundings. If you are indoors, how many doorways are there? Where are they located? How tall are the ceilings? Are there windows, and which direction do they face? Now, think of the exterior of the structure. How did you arrive at this place? Were there sidewalks, hallways, or other pathways? Is the structure part of a larger network of homes or buildings? We could zoom out even further, noting how the structure is situated within the town or broader geographic region. When was the structure erected? Is there a historical context associated with its development or with the land upon which it was built? What is the primary purpose of the structure in which you find yourself? Is it commercial, retail, residential, or academic? How does the structure promote specific activities for which it was intended? These are only some of the questions that we can ask about our surroundings. Too often, we are somnambulists in a well-designed world. But design is rarely coincidental. As I have argued, it is this non-coincidence coupled with the ethical posture of the designer that colors our perceived realities and, by extension, our world. Technological intentionality often overrides human agency – not always because it overpowers us, but because we let it. In their book chapter titled "Designing for meaning: The ethical designer's responsibility," authors Stein and Harper articulate why it is essential for us to remain vigilant over our technologies and our dwellings:

> The objects and environments we design or plan play a role beyond the functional. They have the potential to enhance (or to inhibit) human

> flourishing, to contribute to (or detract from) the meaning of the users' lives, to humanize (or dehumanize) them, and to increase their alienation or help them feel 'at home' in the world.[6]

Architecture is not the same as *building*. The main difference, broadly speaking, is that architecture has agency. Architecture focuses on design, whereas building focuses on construction. And it is this design, aesthetic, and configuration of buildings that "speaks" to us in explicit and implicit terms. British interaction designer and researcher Dan Lockton refers to the "architecture of control"[7] in which our built environments shape our behaviors in strategic ways. He uses the example of how windows in traditional British school classrooms start at a higher elevation on the wall, which allows for natural light to enter the room but prevents students from being distracted. Higher windows restrict students' sightline to the playground area outside. The architect thoughtfully planned for these conditions; the builder installed the windows. And let me be clear: I have nothing against builders. I worked as a marine construction laborer for several years to pay for graduate school. I trudged through the North Carolina swamps and intracoastal marshes, building residential piers, bulkheads, groins, and floating docks. But I was merely following the blueprint plans designed by someone else who had carefully considered things like easement setback lines, tidal changes, wake effects of boats and commercial tankers, the ecological effect of the shadow cast on the wetlands below, and more. The architect designed how these structures would work for the user and for the environment; I was a laborer who just put it all together.

Each and every building and manmade environment was planned by someone – or a group of people – who had vision, intent, objective, and purpose. In his book, *How Architecture Works: A Humanist's Toolkit*, celebrated writer and architect Witold Rybczynski persuades us to interpret our environments and to understand the ethics behind them: "The architect takes a position with respect to history and style, details and structure, and plan and setting...We should open our minds – and our eyes – to the richness of our surroundings. And allow the buildings to speak to us."[8] Relatedly, Christopher Alexander writes about "patterns of events"[9] in his seminal work, *The Timeless Way of Building*, in which he encourages the reader to decipher and decode the redundancies in design – the very same patterns that make us catatonic somnambulists. And in *The Great Indoors: The Surprising Science of How Buildings Shape Our Behavior, Health, and Happiness*, Emily Anthes details how our spaces dictate much of our lives, without us ever fully realizing it. She cites numerous built environment studies that correlate our architectural designs with better literacy rates, physical and psychological health, and even lifespans. Nowhere in her book does the phrase "ethical design" appear, but Anthes consistently refers to what she deems to be "good design," which "expands what's possible. It nudges us in the right direction, supports cultural and organizational change, and allows us to express our values." She concludes "We are products of our environments, but we don't have to be victims of them."[10] Ethical design starts with the designer.

Affordance theory

James J. Gibson's 1979 work, *The Ecological Approach to Visual Perception*, introduced an important concept in design: *affordances*. Specifically, he was musing about the relationship that exists between an animal or organism and its environment, but eventually, by Chapter 8, Gibson arrives at a powerful supposition. He notes that the built environment – and the objects contained within it – consists of implied meanings that are perceived and interpreted by the person (or animal) that interacts with it. He says that, "The *affordances* of the environment are what it *offers* the animal, what it *provides* or *furnishes*, either for good or ill."[11] Further, it is both the physical properties of the environment and its objects (affordances) and our relative understanding of how to use the environment and its objects (perceived affordances) that guide our behaviors. In other words, according to Gibson, an affordance "is equally a fact of the environment and a fact of behavior."[12] The object or the environment doesn't necessarily make you do something, but its characteristics suggest and invite specific interactions. For example, suppose you are walking down the street and encounter a soccer ball lying on the ground. Your brain instantly processes all the potential actions, or *affordances*, that this soccer ball presents to you. Perhaps the most common interaction would be to kick the ball with your foot. (Note that I did not specify in which direction or at whom you were kicking the ball, as this is somewhat irrelevant to the point). The main affordance of this soccer ball – what it offers you to do with it – is to kick it. Although there are other affordances (e.g., the ball could be picked up and thrown, you could jump on top of it, or you could even take a bite out of it), we interpret the ball as something to be kicked. Now, suppose the soccer ball is deflated and caved in. You likely would not kick a soccer ball in this condition; although the affordance of kicking still remains, your *perceived affordance* has changed. You know that kicking a deflated ball will not be the same as kicking an inflated one, so maybe you walk past the ball without any interaction at all. The deflated soccer ball has altered your perception of what it affords the user. Lastly, suppose this soccer ball is realistic-looking but made of concrete. The actual affordance (to kick) technically remains, but the materiality has changed. Kicking a concrete soccer ball seems a highly unlikely action since a reasonable person would deduce that this would result in injury. And if the soccer ball was as large as an SUV or as small as a marble, kicking it would still be a possible yet uncommon way to engage with the object. So, what Gibson suggested is that the physical properties of an object (or environment) coupled with our "psychical"[13] perceptions of how objects are to be used suggest to us how we should interact with it. The way we interact with the soccer ball is guided by a number of factors that can be manipulated through design.

In the field of Science and Technology Studies (STS), we often talk about how objects *mediate* our existence with our surroundings. Understanding this, designers can cleverly arrange features and elements of objects and environments to encourage specific behaviors. Ethical design, then, examines the conditions under which the object or environment was designed and then evaluates whether it has the well-being of its users (and non-users) in mind. For instance, French philosopher Bruno Latour uses the example of speed bumps to illustrate how objects can be "scripted" to prescribe certain behaviors to their users. A school or neighborhood might install speed bumps on their roads to force drivers to slow down; this is much more effective than posting a SLOW sign on the side of the road, which could simply be ignored by drivers. But we would not want to install speed bumps on the road leading to the emergency room entrance at the hospital. Doing so would force drivers (including ambulances) to slow down when time is of the essence and could be quite literally a matter of life and death. Speed bumps are an example of a "forcing function," which shapes behavior through an inescapable action that requires the user to act in a specific way. For example, you must remove your debit card from an ATM before it dispenses cash, as this prevents you from leaving your card behind in the machine. This design is more than just a deterrent; it requires that you perform a specific action before you can proceed. According to the Interaction Design Foundation, forcing functions are meant to "deliberately disrupt the efficient or automatized performance of a task."[14] In other words, force function designs capitalize on perceived affordances to wake the user from their somnambulism. A forcing function might be designed to protect the user (such as with a well-placed speed bump or a childproof twist top on a prescription medicine bottle), but it also can be an unethical strategy (such as with website popups that ask for personal information like an email address before a user is able to view the page). Forcing functions are all around us, and they strip us of our autonomy and control. And while forcing functions can be annoying, so can objects and environments that are simply poorly designed.

Don Norman extended the concept of affordances to the field of interaction design in his 1988 work, *The Psychology of Everyday Things.* In it, Norman critiques the design of everyday objects like light switches, control panels, appliances, and perhaps most famously, doors. As I touched on briefly in Chapter 2, a "Norman door" is a term used to describe a door that gives off ambiguous signaling to its user. The image below includes a "PUSH" sign, but it also has a handle designed to be pulled. So, which action should the user perform?

Figure 3.1. Norman doors fail to clearly signal whether to push or to pull.

An acute observer might note that the door jamb is on the outside of the door slab, requiring the door to be pushed inward rather than pulled outward, but these decisions are often quick and thoughtless, and users are more likely to rely on a trial-and-error method to enter the restroom. Norman uses the term *signifier* to describe the observable cues in design, such as the physical features of the "PUSH" placard and the pull handle in the image above. These signifiers communicate to us the affordances of the object, but when this message is contradictory (such as with a door that has a "PUSH" sign *and* a pull handle), the user is left confused. The distinction between *affordances* and *signifiers* is an important one; the former is the potential actions that one can take, and the latter is the physical properties of the object that communicate these actions to the user.[15]

In recent years, there have been many advancements in the niche market of door-opening. Hands-free mechanisms such as foot pedals and arm pulls can be added to doors to give the user the option to use their foot or forearm to open the door. Furthermore, the COVID-19 pandemic gave rise to the popularity of "sanitary keys," which people could easily carry with them and use to pull open door handles and as a contactless stylus to push buttons, switches, and any other surface which you wanted to avoid touching. These are simple, low-stakes design modifications to ordinary behaviors like door-opening, but they promote well-being through ethical design, as their design and implementation are intended to prevent the spread of harmful viruses and bacteria on restroom doors. In fact, many public restroom areas have upgraded their facilities so that users can attend to their business without ever having to touch a single surface; self-flushing toilets and motion-sensing faucets and hand dryers ensure a sanitary but contactless experience. Some retail stores have redesigned the architectural layout of their restroom area to eliminate entry doors altogether. Users instead enter through a small labyrinth of staggered walls that maintain privacy but allow for touchless entry into the restroom area. When they are well-designed, objects and spaces communicate their affordances to us, whether it is a foot pedal or forearm pull affixed to a door or a complete reimagining of how to enter a restroom area.

Sometimes, however, affordances can challenge conventional norms and our perceived realities. If an object seems disjointed or dysfunctional, it becomes immediately noticeable to us. Consider the surrealist art movement that became popular in the early to mid-part of the twentieth century. Artists like Salvador Dali, Rene Magritte, and Wolfgang Paalen produced works that were absurdly mind-bending and provocative. Dali's painting *The Persistence of*

Memory (1931) is immediately recognizable and found its way into many college dorm room walls for decades to come. The painting shows melted clocks and a fleshy blob with long eyelashes sprawled on the ground, looking vaguely familiar and human-like but clearly foreign and indecipherable. Dali would revel in this craft of what he described as “eye-fooling.”[16] Similarly, Magritte’s 1929 painting titled *The Treachery of Images* features a pipe with a phrase underneath it that reads: “Ceci n’est pas une pipe” (French for “This is not a pipe”), though the object is very clearly a pipe. Magritte’s famous work disassociates our objects from the language that we use to describe them, stating that his intention behind his artwork was to “make the most everyday objects shriek aloud.”[17] And it was Wolfgang Paalen who deviated from painting as a medium and began constructing physical objects, a practice called *assemblage*, which combined (or assembled) various objects and materials into unlikely combinations. His object titled *Nuage Articulé* (1937) is an umbrella that is made from sponge-like material.

Figure 3.2. Wolfgang Paalen’s *Nuage Articulé* (1937) violates our perceived affordances.

Many other twentieth-century surrealists would dabble in assemblage with their own playful designs. Meret Oppenheim was a Swiss artist who exhibited works like *Object* (1936), which was a fur-lined teacup, saucer, and spoon and *Sugar Ring* (1936), which featured a sugar cube set into a gilded silver ring. The objects elicit an internal struggle in the spectator. Although we recognize them as ordinary *things*, we also anticipate how they will most certainly fail; a sponge umbrella won't be of much use in a rainstorm, a fur-lined teacup is virtually unusable, and a sugar cube ring will dissipate over a short period of time. The reason that this art form is so powerful is because it contradicts our language of affordances and usability.

Today, modern artists continue the practice of assemblage, designing objects that are explicitly meant to give us pause. Katerina Kamprani is a Greek architect who designs objects that force the viewer to contemplate their affordances and limitations.[18] The disclaimer on her online portfolio reads: "All the objects you will see in this website are deliberately designed to annoy you."[19] Kamprani's 3D prototypes include objects that contradict basic ergonomic design principles, such as with *Uncomfortable Wine Glass* and *Uncomfortable Chairs*, create blatant inconveniences, such as with *Triple Door*, and commit material transgressions, such as with *Inflatable Paddle*. Kamprani even gives a nod to Paalen's spongey umbrella with her own creation, *Concrete Umbrella*, which is an umbrella made out of – you guessed it – concrete. Similarly, Priscilla Ong is a Singapore-based modern artist who practices assemblage, and her work often confounds viewers by combining foods with everyday objects. *Soft Serve* is a pin-pong paddle with a dollop of whip cream on one side, *Say Cheese* features a single piece of processed cheese as if it were being developed by a Polaroid camera, and *Red Tape* is a clear tape dispenser with wound sour ribbon candy instead of the tape.[20] Intuitively, we recognize the general aesthetic of Kamprani's and Ong's objects, but we quickly predict that each will be problematic in its own way. An inflatable paddle won't propel a kayak or canoe, a concrete umbrella is too heavy to lift, a Polaroid camera could not process a piece of sliced cheese. These common sensical predictions are evidence of affordance theory at work. Kamprani, Ong, and other artists like them cleverly misuse and exploit our perceptions of everyday objects, and in doing so, demonstrate the underlying relationships that we have with material objects, their signifiers, and the affordances that we believe to be true.

I use the example of assemblage and surrealist art to extend a broader point about affordances and the built environment. After all, the philosopher Bas de Boer describes the environment as a "rich landscape of affordances."[21] We are immersed in a world of *things*, of which we are constantly scanning, interpreting, understanding, and envisioning their uses. Our environment is the sum total of these things – products, processes, and systems – and it is inescapable. Therefore, it is extremely critical that we are aware when a design has become corrupted and compromises ethics.

Hostile design

As I have shown, the design of an object plays an integral role in how we perceive and use it. If design is the blueprint for how we should interact with something, whether it be an individual object or the greater built environment that surrounds us, then it is not a stretch to conclude that design informs our behavior. So, what happens when design promotes deliberately unethical, unjust, and discriminatory behaviors against other, more vulnerable populations?

In his book, *Callous Objects: Designs Against the Homeless*, Robert Rosenberger argues the notion that design (or our objects and our environments) can harbor social, political, and moral prejudices, and as a result, we should be more attuned to the "technological guilt" that results from their use. In Chapter 1, I referred to the strategy of *hostile design* as a way of deterring people from engaging in unwanted behaviors. The example given was perhaps the best-known avatar for hostile design: the divided bench. This is a commonly cited example of hostile design – sometimes referred to as *hostile architecture*, *anti-homeless architecture*, or *defensive architecture* – because of the ubiquity of divided benches in urban settings and its unassuming design. Most of us have sat on a divided bench without giving it much of a second thought (let alone feeling guilty for doing so). But the divided bench has a sneakily simple design with an express purpose: to deny the unhoused population a surface for lying down. In fact, there are many variations of hostile design features that accomplish that same goal: divided benches, raised armrests, circular seating, and narrow surface areas, to name a few. Slanted seats, known as "perch benches," are found in a variety of mass transit waiting areas and afford leaning but only for short durations of time. Strategically, they offer travelers more comfort than standing but less coziness than sitting or lying down; perch benches are designed to accommodate the public *just enough*. They do not allow someone to sit comfortably for an extended period of time, and are therefore, exclusive by design.

Figure 3.3. A perch bench affords leaning but not sitting.

Another staple of defensive architecture is the Camden bench. This seemingly innocent slab of concrete affords sitting but in very specific ways. And it is what these benches preclude that makes them hostile. The wavy edges ensure that skateboarders will not attempt to grind the ledge, the uneven surface makes lying down severely uncomfortable, and its sheer weight makes it impervious to theft or relocation to other areas (absent of a crane).

Figure 3.4. A Camden bench in London, England.

The designer of the Camden bench acknowledges that the object is "a pretty defensive piece of furniture" but defends his architecture for being a successful deterrent of anti-social behavior.[22] Critics of hostile architecture are quick to point out that they do not necessarily condone or support "antisocial behaviors" such as rough sleeping outdoors on public architecture, or encourage defamation or drug use via these structures. However, these behaviors have been deemed *antisocial* through a particular lens that might be caused by a power imbalance and not be shared by all; skaters might welcome not criminalize drug use, street artists may celebrate, not condemn graffiti art, and the unhoused might appreciate, not demonize a well-lit and secure space to rest for the night. It's not that architecture should promote or even tolerate destructive behaviors and systemic issues like homelessness, but it certainly shouldn't police it either. Hostile design pushes unwanted populations out of the cityscape by serving as a constant reminder that they are unwelcome in the public sphere. Furthermore, these designs act as yet another form of policing on marginalized groups of people who already lack a voice. And the designs are getting even more aggressive.

Anti-homeless spikes often blend into the urban landscape as ornamental decorations on ledges and flat surfaces, but their intent is far more combative. Installation of spiked ledges signals to the passerby that sitting or lying is prohibited, but more importantly, we understand (through perceived affordances) that sitting or lying on a spiked ledge would inflict pain. In this sense, the spiked ledge goes beyond the deterrence of unwanted behavior and threatens corporal punishment if ignored.

Figure 3.5. Spiked ledges prevent unwanted sitting.

Spiked ledges are a rather special genre of hostile design because they are the embodiment of *hostility* – both metaphorically and physically. The "Pay & Sit" bench was created by Fabian Brunsing, a German sculptor, as part of an exhibition of street furniture. The park bench features metal spikes on the horizontal slats where a user would sit (or lie down). When the user inserts €0.50 (about .70 cents US) into the coin slot, the spikes automatically retract into the bench. After about one minute, an alarm sounds, giving the sitter time to either buy more time or evacuate the bench before the spikes emerge again.[23] Brunsing's design is a satirical commentary on the commodification of public amenities and the alienation of marginalized groups. While most are appalled at the notion of paying to sit in a public space, the design was adopted by city planners and installed in Yantai Park in Shandong, China. Sometimes, life imitates art.

A Google image search of the phrases "hostile design" and "hostile architecture" reveals pages and pages of creatively barbaric tactics used in architecture to enforce laws against loitering and rough sleeping, some of which include shards of glass, jagged rocks, and barbed wire embedded into concrete to deter people and animals. Undoubtedly, these unethical design features work, but at what cost? Rosenberger elaborates on why spiked ledges are egregious violations of our ethical standards:

> [S]pikes are forthcoming about their intentions... If your moral reasoning and intuition bring you to recognize that designs like the antihomeless spikes are no way to treat fellow citizens and fellow human beings, then it turns out that you also hold an additional moral position: you should be additionally against the larger and less visible panoply of laws and design that target the unhoused.[24]

It is easier (and cheaper) to install a divided bench or spiked ledges rather than to address the complex, nuanced issues of mental health and homelessness. A business owner might turn to hostile design to prevent loitering or rough sleeping around the premises of his property, but in doing so, he also is staking a claim against those populations. Hostile architecture forces us to reckon with our own ethical posture. Would you object to a judge sentencing a jaywalker to life in prison? And if so, why? The answer is likely because the punishment doesn't fit the crime. Similarly, hostile designs like Camden benches and spiked ledges – which are designed specifically for discomfort and even pain – seem like outsized punishments for victimless, antisocial behavior. Remember, homelessness is not a crime.

Sometimes, hostile design is not always visible. Los Angeles County has the largest unhoused population in the United States, with roughly 69,000 individuals at the time of this writing. At the Westlake/Macarthur Metro Station, which is a bustling hub for public transportation as well as for many unhoused individuals, there has been a noticeable increase in the volume of classical music being played in the terminal. City officials describe the excessively loud audio as a pilot program to make the subway station more enjoyable for commuters, but unhoused individuals see it as a deliberate strategy to use audio warfare to displace them from the Metro. One unhoused person interviewed by an NPR reporter was quoted as saying, "The city better work for everybody. That means the rich, the poor, the homeless. There's got to be a place for everybody."[25] It turns out that audio can be quite an effective tool in defensive design. 7-Eleven convenience stores in various locations throughout the United States and Canada have been reported as playing classical music at excessive volumes, and a downtown area in Palm Beach, Florida, blasts children's songs like "Baby Shark" and "Raining Tacos" to deter rough sleepers.

The Mosquito Anti-Loitering Alarm system is a self-described "vandal-resistant powerhouse."[26] The device emits a high-frequency noise that is designed to irritate anyone within earshot, presumably protecting the property by fending off unwanted visitors. A key specification of the $600 "anti-social" device is that the sound can be programmed to a specific frequency, 17 KHz, which is audible to only those under the age of 25. According to the website, "The deployment of Mosquito devices gives communities respite in cases of acute anti-social behaviour and youth nuisance... The tone is highly annoying and is used to prevent and disperse teenagers congregating outside property and business premises."[27] The design literally discriminates by age, using the natural degradation of hearing (a condition called presbycusis) as its filter. The Mosquito is more than a security system; it is an ageist tool designed to thwart "youngsters." The website suggests the following locations as being suitable for installing the Mosquito anti-loitering alarm: convenience stores, car parks, subways and rail stations, schools during evenings and weekends, public parks after dark, and more.[28] Whether it is classical music or high-pitched noise frequencies, architecture can easily become weaponized against powerless groups.

Defensive architecture does not merely deter unwanted or antisocial behavior; it creates a hostile environment. As a critic of hostile design, I fully accept that certain types of behaviors may be inappropriate and unwanted by others; to be clear, I am not defending the right to vandalize or misappropriate public spaces or structures. I am pointing out that hostile designs presume the worst in everyone, target and discriminate against specific groups, and use design to antagonize and marginalize people rather than provide helpful solutions. The built environment should promote the welfare and well-being of all people, not be a mechanism for policing behavior.

In recent years, there have been efforts to combat homelessness in urban areas by using a housing-first approach that provides the unhoused with a clustered community of shelters (sometimes referred to as micro-houses, micro-living, tiny house villages). Some places provide mental health, medical, and education services in addition to basic housing. Medicine Hat is a city of approximately 63,000 people located in Alberta, Canada. In 2009, its elected officials developed a plan to end homelessness (as defined by having no more than three unhoused individuals experiencing homelessness for three consecutive months). Using a housing-first approach, the city declared itself as having "functionally zero"[29] homelessness in less than five years. Building micro-houses and communities for the unhoused is not cheap, but neither is hostile architecture. There is a significant financial cost for installing or retrofitting hostile designs throughout any given cityscape; in addition, when homelessness and rough sleeping are criminalized, there are other costs associated with increased law enforcement presence and subsequently arresting, arraigning, and incarcerating unhoused individuals. The online publication, *The Homeless Voice*, estimates that it costs roughly $31,000 to criminalize one unhoused person.[30]

Hostile design has become a symptom of broader cultural issues, as it serves as a proxy for our societal biases that favor incarceration over rehabilitation, criminalization over reform, judgment over empathy, and the individual over the common good. Often, these designs blend into the nooks and crannies of the urban landscape, effectively serving their purpose yet avoiding scrutiny from most. But it doesn't have to be this way. We can build a good architecture that fosters a humane society, not a hostile one. We can support ethical designs that prioritize wellbeing. We should incorporate more resources for the unhoused population rather than install spiked ledges on every flat surface. Build a skatepark instead of criminalizing skateboarding. Designate spaces for public art display. Design spaces for safe gatherings rather than pump high-pitched noises into the environment to shoo away the youth. Architecture should be designed for inclusivity, not othering.

Infrastructure

There is a curious story that has been whispered among circles of engineers and designers for decades now. The telling of the story has to do with a man named Robert Moses, his racist and bigoted views, and the urban infrastructure that he championed throughout his career as an urban planner in New York City. Robert Caro was the first to report on the story in his 1974 Pulitzer Prize-winning biography, *The Power Broker: Robert Moses and the Fall of New York,* a 1,336-page analysis of the prolific designer and builder of roads and parkways, green spaces, and landmark buildings in the New York metro area throughout the early and mid-twentieth century. Caro repeatedly refers to Moses as the "master-builder" given that he was the architect for the central nervous system of the most populated city in the United States. According to the author:

> Robert Moses shaped New York. Physically, any map of the city proves it. The very shoreline of metropolis was different before Robert Moses came to power. He rammed bulkheads of steel deep into the muck beneath rivers and harbors and crammed into the space between bulkheads and shore immensities of earth and stone, shale and cement, that hardened into fifteen thousand acres of new land and thus altered the physical boundaries of the city.[31]

While it is indisputable that Moses left his mark on the terrain, his motivations are less obvious. Caro asserts in his meticulously documented biography that Moses began his career as a democratic idealist but ultimately became corrupted by his pursuit of power, which he ultimately obtained and then wielded over influential politicians, business tycoons, and prominent social figures. And it is at this point in his career, according to his biographer, that Moses began to infuse his personal discrimination into his public-facing designs.

Of the many designs and builds that spanned Moses's career, the Southern State Parkway is his most scrutinized. This 25-mile stretch of road traverses across Long Island, NY and provides a scenic getaway to Jones Beach, a recreational beach area that was used predominantly by Whites. The problem? Many of the overpass bridges on the Southern State and Grand Central Parkways (as designed by Moses) are uncharacteristically low and physically prevent public transit from passing under them, thereby favoring automobiles and denying mass transportation vehicles like busses from reaching Jones Beach. At the time – in the 1920's – public transportation served poorer, disenfranchised minority groups, primarily. Moses's racist views had informed his design to keep minorities out of Jones Beach, or at least this was Caro's conclusion. The biographer wanted to prove that Moses's intentions were racially driven, so Caro and his wife even went as far as to spend an entire day at the entrance to Jones Beach, tallying visitors according to their race. Their small-sampled, unscientific experiment did reveal a disproportionate number of White beachgoers, though this does not necessarily indicate that Moses successfully executed his racist infrastructure.[32] This allegation would require more credibility.

The Power Broker relies heavily on the testimony of Sidney Shapiro, the former chief engineer of the Long Island State Park Commission and close colleague of Robert Moses. Shapiro provided more than 100 hours of interviews with Robert Caro on the condition that nothing be published until after his death. Caro complied. Shapiro provided inflammatory statements that asserted Moses's discriminatory designs in infrastructure projects such as public swimming pools. In *The Power Broker*, Shapiro alleges that Moses deliberately lowered water temperatures in public swimming pools in racially integrated neighborhoods because of his misguided belief that African-Americans did not like to swim in cold water. Shapiro also corroborated the accusation that Moses intentionally built low overpasses to deny public transportation riders access to Jones Beach. Testing the theory that Moses designed low overpasses as a deterrent, one reporter from *Bloomberg* compared every one of Moses's overpass heights to contemporary parkway designs throughout the greater New York area and found that the clearance heights were indeed significantly lower on parkways that were designed and built by Moses.[33] This seems to confirm Moses's racist intentions, as told by his chief engineer, Shapiro, who once recalled Moses as having used the term "dirty" to describe minorities.[34] Additionally, Moses's legacy remains complicated to this day, with some seeing him as a bigot and others praising him for his prolific contributions to urban planning and design.

Despite the physical evidence of extremely low overhead parkways and the firsthand accounts that offer up his intentions, many people have tried to debunk the claims that Moses used the design of his overpasses to enforce his racist beliefs. For instance, some have argued that low overpasses and restrictive bridges were aesthetic choices that *needed* to be made in order to preserve the beauty of the parkway; others have argued that the uncommonly low thoroughfares were due to structural engineering limitations and building costs. Certainly, there is some validity to these claims, yet it is disingenuous to suggest that racism was not a contributing factor. In 2021, nearly 100 years after construction on Robert Moses's parkway began, Transportation Secretary Pete Buttigieg made the following remarks:

> [I]f a highway was built for the purpose of dividing a White and a Black neighborhood or if an underpass was constructed such that a bus carrying mostly Black and Puerto Rican kids to a beach — or that would've been — in New York was — was designed too low for it to pass by, that that obviously reflects racism that went into those design choices.[35]

This seemingly reasonable conclusion drew ire from right-wing politicians like Ted Cruz (R-TX), who sardonically responded to Buttigieg, saying, "The roads are racist. We must get rid of roads," and from conservative talk show pundits, such as Tucker Carlson, who said flatly, "Inanimate objects, like roads, can't be racist."[36] (Not to be overly dramatic, but I find this last statement to be the antithesis of this entire book.) We will likely never reach a conclusion on Robert Moses's design intentions, but most accounts describe him as being a generally unpleasant individual, so it is not too difficult to infer that he also espoused racist sentiments during a period of extreme racial prejudice and inequality in America. Regardless, it is not the goal of this book to prove whether Robert Moses was an unabashed racist or not; it is to show that inanimate objects, processes, systems, and institutions absolutely can harbor racism, and we should be mindful of the hidden ethics of the built environment.

The first part of Buttigieg's comments – a "highway built for the purpose of dividing a White and a Black neighborhood" – is a reference to yet another historical example of how infrastructure has been used as a segregationist tool. In his book, *Race and the Shaping of Twentieth-Century Atlanta*, Ronald Bayor writes that in the 1940s, Atlantan mayor William Hartsfield deliberately planned for a major highway, I-20 West, to physically segregate neighborhoods along the lines of race. Bayor reported on a 1960 report from the Atlanta Bureau of Planning which stated that the proposed highway "would be the boundary between the White and Negro communities."[37] This would not be the last attempt to racially divide groups in Atlanta. In 1962, Mayor Ivan Allen Jr. approved the measure to erect barricades on Peyton and Harlen Roads, which effectively prevented vehicles from traveling into the Cascade Heights neighborhood (a wealthy White suburban hamlet) from the nearby neighborhoods that were predominantly poor and Black. Days after they were installed, the *Atlanta Journal-Constitution* reported that the road barriers were an explicitly racist "buffer ordinance, aimed at stemming the expansion of

Negro families into previously all-White neighborhoods."[38] The barriers, which also blocked fire and emergency services from entering surrounding Black neighborhoods, would become known as "Atlanta's Berlin Wall" and ultimately were ruled to be unconstitutional by a Superior Court Judge who ordered their removal due to pressure from local Civil Rights leaders. The legal victory to remove Atlanta's Berlin Wall was more of an exception than a rule, however. In most cases, disenfranchised communities stand little chance of opposing major construction projects that are government-funded.

The practice of *urban renewal* is a clever euphemism for the forced migration of underprivileged populations for the benefit of redeveloping the land for public housing and other economic investments. In the United States, this strategy was legislated as part of President Harry Truman's Fair Deal in the 1949 Housing Act, which provided financial subsidies to local governments for "planning, code enforcement, and the rehabilitation of structures and neighborhoods."[39] The redevelopment efforts were referred to as *slum clearance* and disproportionately affected residents in poor neighborhoods who had little or no agency. Modern-day slum clearance continues around the world. Dharavi in Mumbai, India, is considered to be the world's largest slum, and in 2023, Indian billionaire Gautam Adani submitted a $619 million bid to redevelop the area. The project would displace more than 1 million people and has been described as "the world's largest urban renewal scheme."[40] The United Kingdom waged its own slum clearance efforts throughout the twentieth century, ousting millions of people from their homes and communities. Often, there is an undercurrent of racism and cultural insensitivity that drives urban renewal programs, as evidenced by this 1933 article in *The New York Times* that begins with the sentence, "For generations the slums of the East Side have been a disgrace to New York -- and to America."[41] When entire neighborhoods are deemed to be "slums" or "blighted" and earmarked for redevelopment, historical and cultural significance can be lost.

The Greenwood District in Tulsa, Oklahoma, was named "America's Black Wall Street" in the early 1900s because of its economic prosperity and the successful business acumen of its residents, nearly all of whom were African American. But in 1921, racial tensions over the land use resulted in two days of death and destruction. Hundreds of Greenwood residents were murdered, their bodies dumped into mass graves, as 35 city blocks of Black Wall Street were burned and destroyed by White antagonists. Two days after the massacre, the mayor rezoned the Greenwood District for industrial business, and 8,000-10,000 people were left homeless.[42] It would take decades for the Greenwood

District to be rebuilt, particularly because the newly adopted zoning laws made it unaffordable, requiring the use of more expensive building materials for new construction. For those Black Americans who returned and reinvested in the area, the Tulsa Race Massacre would haunt them for the rest of their lives. But the Greenwood District would make a comeback during the 1940s, 50s, and 60s with the re-emergence of many Black-owned businesses. Then, the Federal-Aid Highway Acts of 1965 and 1968 proposed and built interstate highways that sliced through the Greenwood District. An article in the *Smithsonian Magazine* described it this way: "What the city could not steal in 1921, it systematically paved over 50 years later."[43] Across the continental United States, the federal legislation touted the new highway system as a boon for interstate commerce, but it often was routed through historically Black areas. In one example, the state highway director of Montgomery, AL (who also was a prominent member of the Ku Klux Klan) deliberately planned Interstate 85 to run through a Black neighborhood home to local Civil Rights leaders rather than through nearby uninhabited land.[44]

The National Institute on Minority Health and Health Disparities defines *structural racism* as "macro-level conditions that limit opportunities, resources, power, and well-being of individuals and populations based on race/ethnicity and other statuses."[45] With respect to Tucker Carlson's objection that "roads can't be racist," structural racism practices such as urban renewal and slum clearance show us that roads most certainly can promote racist ideology and create racial inequality through segregation and erasure. On one hand, paved roads and highways are inanimate objects that do not hurl racial epithets or ethnic slurs at passersby. On the other hand, they concretize (both figuratively and literally) racist beliefs and the total disregard for the rights and well-being of vulnerable and marginalized populations.

Our infrastructure rests on the intentions, motivations, and ethics of others that came before us. We drive on streets that were laid at the expense of entire neighborhoods and communities. We navigate cities that continue to honor these figures. This reminds me of a passage from Henry David Thoreau's *Walden*, which reads:

> We do not ride on the railroad; it rides upon us. Did you ever think what those sleepers are that underlie the railroad? Each one is a man, an Irishman, or a Yankee man. The rails are laid on them, and they are covered with sand, and the cars run smoothly over them. They are sound sleepers, I assure you. And every few years a new lot is laid down and

> run over; so that, if some have the pleasure of riding on a rail, others have the misfortune to be ridden upon.[46]

In 2016, Robert Moses Parkway was officially renamed as Niagara Scenic Parkway, but there are still numerous references to Moses throughout the New York area, including beaches, schools, parks, and even statues. The Hartfield-Jackson Atlanta International Airport invokes Mayor Hartfield, who physically segregated Black and White neighborhoods. In 2013, Tulsa City Council voted to change the name of its Brady Street (after Tate Brady, a KKK member) to Matthew Brady Street (after a Civil War photographer). But after public outcry from city residents, the name was changed once again to Reconciliation Way. Racial bigotry still persists on our roads. Many neighborhoods – particularly in the Southern United States – are named thematically to honor the Confederacy. If you take the time to look closely, you'll find that our built environment tells the story of those who have been "ridden upon." And in doing so, our infrastructure galvanizes our ethics and values of who we are and what we are willing to forget.

Case study: Amager Bakke

Eco buildings – as the name suggests – are structures that are environmentally conscious. This could mean that the building materials are ethically sourced or sustainable, or that the structure itself is self-sustaining or maximizes energy efficiency, or that the overall architectural design benefits the common good. In Copenhagen, Denmark, for example, architects from the Bjarke Ingels Group (BIG) turned a $590 million waste-to-energy powerplant into a major tourist attraction and recreation destination. Amager Bakke (also known locally as CopenHill) opened in 2017 as a municipal powerplant capable of processing an estimated 440,000 tons of solid waste into clean energy that can power more than 30,000 nearby homes with electricity and 72,000 homes with heating and cooling throughout Copenhagen and its surrounding boroughs.[47] But what makes this powerplant unique is not its environmentally friendly waste processing and conversion; the 410' building was designed with urban recreation in mind.

The 441,000 sq ft. building, which has been described as the "cleanest and most energy-efficient waste-to-energy plant in the entire world," features a rooftop park open to the public, trails for hiking and for mountain biking, a 1,480' artificial ski slope, and a 279' artificial climbing and rappelling wall (the largest in the world) – all situated on top of a massive solid-waste incinerator.

Figure 3.6. Amager Bakke is Copenhagen's largest waste-to-energy plant and ski slope.

In his book, *The Architecture of Ethics*, Professor Thomas Fisher describes how much of our built environment mostly follows "hedonistic utilitarian" ethics, where profit maximization and familiarity often outrank aesthetics and the common good. For example, a builder might propose a housing complex that maximizes the number of family units it can accommodate or use the most economical building materials and fixtures that would yield the most profit. Alternatively, an "ideals utilitarian" would consider "intellectual, aesthetic, or even spiritual values, and not just those related to physical pleasure or economic profit."[48] It cost an additional $14 million in private funding to make Amager Bakke into a recreational landmark in a region without naturally formed mountains. A hedonistic utilitarian might ask whether this added cost was worth it, given that most of the amenities are free to use, and the ski slope only costs $33 per person. But an idealistic utilitarian would recognize its immeasurable value added to the local community.

Copenhagen, which has a population of more than half a million people, is especially invested in eco-designs like Amager Bakke, having pledged to become the world's first carbon-neutral city by the year 2025 while still recognizing how its ethically designed architecture can benefit its citizens. In 2023, the city was named the World Architecture Capital by the United Nations Educational, Scientific, and Cultural Organization (UNESCO) because of its human-centered infrastructure. Other notable architecture in Copenhagen includes bicyclist and pedestrian-friendly roadways and bridges, wide-open public parks and playgrounds, accessible waterfronts for recreational swimming and boating, and visually stimulating buildings. Perhaps this contributes to why Denmark is consistently ranked as one of the happiest countries in the world, according to the World Happiness Report.[49]

Eco-building exists all around the globe. One Angel Square in Manchester, UK; the Museum of Tomorrow in Rio de Janeiro, Brazil; Torre Reforma in Mexico City, Mexico; and the Bosco Verticale in Milan, Italy, all exemplify cutting-edge design in sustainability and green building technologies. In order to be considered an "eco-building," structures must meet rigorous conditions set forth by a consortium of evaluative bodies and organizations that certify eco and green building practices, including the Leadership in Energy and Environmental Design (LEED), the Israeli Green Building Standard #5281, the Building Research Establishment Environmental Assessment Method (BREEAM), World Green Building Council (WorldGBC), the UK Green Building Council, German Sustainable Building Council, to name just a few.

The most ambitious eco-building project might just be NEOM – a region in Northwest Saudi Arabia that will include several large-scale mega-construction projects. The Crown Prince and Prime Minister, Mohammed bin Salman has dedicated financial resources from the Public Investment Fund (PIF), considered to be the largest sovereign wealth fund in the world, valued at more than $750 billion. The word NEOM is a combination of the prefix NEO-, meaning new, and the Arabic word Mustaqbal, meaning "future." A promotional video on NEOM's website describes the region as "an entirely new model for sustainable living."[50] The building portfolio includes projects like Oxagon, a re-imagined port city; Trojena, a mountainous recreation area; Sindalah, an engineered island retreat; and perhaps the most ambitious and innovative project, The Line.

The Line is a 170km (~100 miles) long linear city that will be 200m (~220 yards) across and has 500m (~1,640')-high mirrored walls on each side. There will be no roads or cars, which means there will be no emissions. NEOM project leaders predict The Line will be home to 9 million residents, though some have pointed out that this equates to roughly the "density of 265,000 people per square kilometer – ten times denser than Manhattan and four times denser than the inner districts of Manila, currently estimated to be the densest urban neighborhoods on Earth."[51] In addition to the plethora of engineering and environmental obstacles that NEOM faces, it is planned to run directly through tribal lands. The Huwaitat are one tribe that inhabits this region of the desert, so the Saudi government has offered financial incentives to natives who are willing to relocate. For those who refuse to leave, harsh penalties may be imposed; in 2020, as construction on The Line began, three men from the Huwaitat tribe were arrested and sentenced to death for refusing to evacuate their homeland, according to a human rights watch group.[52]

There are many other experimental sustainable city initiatives across the world that will encounter similar ethical challenges. There is The Sustainable City, located in Dubai, United Arab Emirates; the Floating City in the Maldives; Biodiversity in Malaysia; and Telosa in the United States, all of which are conceptual visions of living sustainably in futuristic worlds. As we continue to reinvent the ways that we build our cities, we might remember the lessons learned from past "urban renewal" projects to ensure that we not only marvel at the new innovations in architecture but also preserve our ethics that become enshrined within. To this extent, we might also consider the degree to which designers, engineers, and builders are responsible for the development of public spaces that are more ideals-based rather than strictly hedonistic.

Chapter 4

Nature

Every technology is an expression of human will.[1]
Nicholas Carr

Design is everywhere. We cannot escape it. The noun form of *design* refers to the way that something has been made or arranged for a particular use. The verb form *to design* means to make decisions about something being built or created. We often equate the term *design* with human effort; for instance, the screen or the printed paper on which you are reading has been designed by humans for the express purpose of viewing alphabetic font that can then be internalized, processed, and understood by an English-speaking reader. In other words, this book was not a natural occurrence nor a random happenstance; it is the output of the author's effort. But, design is not limited to humans. Suppose you were to embark on a voyage to reach the most remote place on earth. Many geographers would point you in the direction of Point Nemo: a speck of an island in the center of the Pacific Ocean, more than 2,600km (1600 mi) from the nearest land. The destination is so remote that international space agencies use it to intentionally crash their de-orbited earthbound spacecraft on it, to avoid any potential harm to other humans. At first glance, Point Nemo is a barren wasteland – just rocky formations jutting out of an otherwise oceanic desert. Yet even the official 'middle of nowhere' is teeming with design. Its oceanic currents and high winds wrap around the steep cliffsides, carving their way through and eroding the geological terrain to find the fastest route for air and water flow. Its ecosystem is home to exotic microorganisms that follow the same natural selection processes that Charles Darwin observed in his landmark book, *On the Origin of Species*. The natural laws of physics and the complex architectures of life still function the same in Point Nemo as they do in Manila (the world's most densely populated city).

As this chapter will show, human-centered design and the natural environment often are at odds with one another. Just as we modify our built environment to suit our needs and our ethics, so too can we reshape our natural environments. Macro-design issues like the ones discussed in this chapter highlight the ways in which the decisions of a few can impact the masses. As a result, the relatively new branch of *environmental ethics* studies how ethical behaviors impact the symbiosis between humans and their built and natural environments. It centers on discussions of policy, social attitudes, and ethical responsibility to our ecology as it pertains to us and to future generations.

The physical manifestation of a biological pattern in living and non-living things is the essence of design: *modification for a particular use.* There are countless examples of evolutionary designs in living beings. For example, penguins originated from seabirds nearly 60 million years ago.[2] But, the iterative exploitation of traits and abilities has left them flightless today. Across generations, their genetic mutation favored their need to dive underwater for sustenance, trading the ability to fly for the expert skill of swimming and diving. Darwin's scientific work clearly affirms design in evolutionary biology. But, scientists and researchers now believe these evolutionary processes influence non-living systems as well. Dubbed the "missing law of nature," scientists have long believed that naturally occurring, non-living systems such as hurricanes and celestial star formation evolve following a complex pattern of iteration that seeks to maximize efficiency. An article published in the *Proceedings of the National Academy of Sciences* introduced a new macroscopic law of nature, which asserts that evolution is a key driver of the natural world, not just living organisms.[3]

These design processes in living and non-living systems are sometimes misconstrued as evidence of *intelligent design* – the belief that traits and the order of things are the result of a forward-thinking being (such as God) who is executing a planned design to make things the way things are. However, this approach confuses *cause* and *effect.* One example, as illustrated by The Ethics Centre, is that an intelligent designer gave giraffes long necks so that they could access food sources higher up on trees, providing them with an abundance of food. Conversely, the evolutionary biology perspective is that giraffes with shorter necks were unable to access this food source; thus, they were replaced by giraffes with longer necks due to their competitive advantage. Intelligent design assumes that an invisible force is making decisions about the way things are. One of the many issues with this approach, however, is that intelligent design credits an omnipotent god with the decision-making, presumably one who promotes both ethical and unethical designs, where some organisms thrive and not others. Unlike evolutionary biology, intelligent design is not a scientific theory because it is unable to be tested.

Critically, I support neither the intelligent design view nor the guided evolution view that natural evolution is assisted by a mysterious figure. This theological view is referred to as *teleology*. The teleological argument attempts to explain the natural world through the lens of guided design. But, a paper in the journal *Evolution: Education and Outreach* describes two types of teleological misconceptions: design-based and natural-based.[4] The former prescribes an intentional, purposeful rationale for why something has evolved; the latter assumes something has evolved out of its predecessor for its own well-being, regardless of intentionality. The teleological perspective starts with the conclusion that there is a god and then uses the natural world as its irrefutable evidence – a type of backward causation. Furthermore, the teleological view would then require that we consider the ethics of the designer (God) if he or she is leveraging natural evolution that clearly benefits some and hurts others. That said, the intent of this chapter is not to prove or disprove the existence of God, but we must establish a baseline that design in nature happens without a designer behind the scenes pulling strings. Nature evolves and adapts out of the need for efficiency and survival, not out of personal motive or vindictiveness. Some religious groups even explain natural disasters in terms of God's wrath. A poll conducted by the Public Religion Research Institute (PRRI) showed that most Americans (56%) believe that God controls everything that happens in the world. To this extent, 4 out of 10 Americans surveyed believe that natural disasters are a "sign from God," and 3 out of 10 believe that God uses natural disasters to punish countries and their citizens.[5] But hurricanes are not sentient, nor are they weapons of war; they are naturally occurring. These statistics might help explain why there has been a decrease in the number of people who affiliate with religious groups who view climate change as an impending crisis.[6]

The reality is that we, as humans, can affect change in nature – in good and bad ways. It is imperative to understand not only that design is all around us by virtue of our built and natural environments but also *why* something has been designed that way and whether we should interfere with it in the first place. Sometimes, an ethical design is to abstain from design altogether.

Climate change

There is order in the seemingly chaotic randomness of nature. This is the thesis of Adrian Bejan who has proposed the *Constructal Law*, which asserts that animate and inanimate designs are harmonious; this is evident throughout designs in nature and in human activity. Constructal law does not imply the teleological perspective that a higher order (God) oversees this intelligent design. Instead, Bejan's law rests on the idea of *flow* – everything is in pursuit of an enhanced flow system, whether it be a river winding through the Amazon or the wealth of a business mogul. When *flow* is enhanced, the *design* emerges. This design is both natural (inanimate) and unnatural (animate), and it is a symptom, not a haphazard occurrence. Randomness, then, is not random at all, just well-designed. Or put even more simply, "Design is flow."[7] According to Bejan, constructal law is "how things *should be*."[8] Although constructal law has its merits – there is, undoubtedly, ordered chaos in nature – it endorses a classical mistake in philosophy: the is-ought fallacy. This view makes the assumption that because things are the way that they are, they are meant to be that way, and this could result in apathy, indifference, or a general sense of helplessness. One example of this is the global climate crisis that we currently face.

Regardless of your political affiliation or personal views on climate change, the data do not lie: The Earth's temperature is warming, sea levels are rising, greenhouse gas emissions are increasing, and non-renewable resources are becoming scarcer. However, it is not the goal of this book to convince climate change deniers that climate change is real. Plenty of data-driven books and articles have been written on the subject that provide overwhelming and irrefutable evidence that this is the case. As climate science is not my expertise, I am not here to convince you that humans are responsible for accelerating Earth's climate change. Instead, I argue that the climate crisis is a design problem. And because the environmental effects of climate change impact the wellbeing of all of us and of our future generations, I see the climate crisis as being an *ethical* design problem, specifically. We must be mindful of the products, processes, and systems that we adopt and their lasting environmental impacts. Additionally, we must continually evaluate our behaviors to determine whether they are ethical. Saying that the climate crisis is a design problem implies that there is a solution, but this is misleading. What I mean instead is that there are innumerable large-scale and small-scale solutions that can be implemented using ethical design practices, and we will need to entertain as many as we can to adjust to a changing climate.

Large-scale design problems require large-scale solutions. In many engineering and architectural schools, students are being taught to think not just as builders and designers, but as environmentalists. The MIT School of Architecture and Planning, for example, has developed a Climate Action Plan, which articulates a commitment to the pursuit of net-zero cities that use green design and construction practices, sustainable and ethically sourced building materials, and renewable energy systems. It turns out that the built environment is a major contributor to global greenhouse gas emissions – a central factor in the rapid acceleration of climate change. According to the United Nations, three-quarters of the built environment is yet to exist and will be built by the year 2050.[9] And the independent organization Architecture 2030 estimates that between the years 2020 and 2060, the world is on pace to build the equivalent of one New York City, per month, for 40 years.[10] This is significant because the built environment, which includes the construction and maintenance of buildings and infrastructures, is responsible for more annual global CO_2 emissions than vehicles and consumer products.[11] The embodied carbon – which is the emission outputs from the extraction, manufacturing, transportation, and disposal of raw building materials – accounts for more than 40% of greenhouse gases.[12] Our current building practices are outdated and compound our environmental woes.

Architectural and engineering ethics demand that builders prioritize environmentally friendly materials, which might be slightly more costly but would immediately lower a building's CO_2 gas emissions by 20%, and use methods of prefabrication and offsite construction, which could lower emission outputs by another 14%.[13] Other climate-friendly building practices include preserving and protecting natural landscapes and animal habitats, optimizing logistics during the building phase, committing to low or zero-carbon emission systems and materials, and planning the built environment for climate resilience. Of course, many of these options are not cost-efficient and require that the engineer, architect, or builder make ethical decisions that might impact their bottom line or compromise the aesthetics of the project. Recognizing this, many professional organizations have created climate pledges, such as the AIA 2030 Commitment (U.S.) and the Five Presidents Pledge (UK and Ireland), and countries have made promises towards more sustainable development, as outlined in the Paris Agreement and the China Accord. But meaningful change might just begin with the next generation. A collective of students has formed the group Architecture Education Declares, which calls for an overhaul of architecture education to address ecological, social justice, and climate issues within the discipline. Still, even with total cooperation, we must face the irreversible damage that we already have caused this planet, prompting some freethinkers to propose radical solutions.

Geoengineering is one such area of theoretical research that advocates for large-scale weather modification projects to counteract the warming of the planet that is due to the greenhouse gases and carbon emissions that trap sunlight. The basic gist of geoengineering is that we might mitigate global warming in either (or both) of the following two ways: by reflecting sunlight back into space (a process called Solar Radiation Management, or SRM) and by removing CO_2 from the atmosphere (a process called Carbon Dioxide Removal, or CDR). In his book, *Unsettled*, author Steven Koonin describes various SRM projects that include the brightening of land and sea by using genetically modified crops, white roofs on built structures, and adding microbubbles to the oceans to reflect sunlight.[14] Another idea – one which seems overly simplistic – is the solar umbrella. Just as a beachgoer sets up an umbrella on the beach to protect herself from harmful UV rays, we could launch a gigantic umbrella into space to shield the planet from sunlight. Barring the crippling costs and obvious logistical challenges that come with what would be humanity's largest construction project ever undertaken, the Natural Academy of Sciences believes it to be a plausible theory. Then there is Carbon Dioxide Removal (CDR). Instead of blocking or reflecting unwanted sunlight from the Earth's surface, we could remove the aerosols in the atmosphere that prevent sunlight from escaping. However, the biggest issue with CDR is the cost to capture CO_2, and then there is the issue of where to store the captured CO_2. The simplest way to implement CDR is to increase the amount of photosynthesis from natural plants, i.e., reforestation, but this likely would not offset human activity. And, as Koonin points out, CDR does nothing to move towards behaving more ethically and only undoes human contribution to climate change. Each of these propositions carries with it unintended side effects that could be far worse than the current rate of global warming that we are witnessing. Additionally, Koonin wonders about the ethics of SRM and CDR, supposing that "a single small nation, a sub national organization, or even a wealthy individual could just do it [SRM]. How would the world respond in that case?"[15] But this may be close to becoming a reality.

Weather engineering is one example of how we already manipulate nature. The Beijing Weather Modification Office in China is an official branch of government that is tasked with controlling weather in and around Beijing. Since the 1990s, the Office has declared that it has been successful in producing snow for New Year's Day celebrations and in creating rainfall to combat severe droughts by injecting "water retaining agents"[16] into clouds to reduce levels of evaporation. The process, known as cloud-seeding, also can be reversed to prevent unwanted rain, as Chinese authorities demonstrated during the opening ceremonies of the 2008 Beijing Olympic Games. The Office used military aircraft and artillery launch sites to infuse nearby clouds with liquid nitrogen and silver iodide, which shrinks water droplets within the cloud and reduces the chance of precipitation.[17] There is some debate over the efficacy of such efforts, but it is indisputable that China routinely engages in weather modification. In Russia, cloud-seeding has been used to tame wildfires in Siberia and to ensure dry weather for festivities in its capital, Moscow. In the United States, at least eight states employ could-seeding efforts to overcome droughts.[18] Just as humans divert rivers and streams to supply populated areas with water, we also modify weather to suit our needs. Weather engineering is not necessarily unethical, as long as it is done with transparency. Although, ethical issues do arise when weather engineering benefits one group and deprives another, when it is used for nefarious or militaristic purposes, or when it has unintended consequences that threaten the wellbeing of others. Since the amount of water on Earth does not change, we might ask whether it is ethical to re-route (or prevent) precipitation to (or away from) a more affluent or populated area over another. Similarly, it would not be ethical to weaponize weather events as a tool of war, such as strategically creating flooding in areas or causing drought-like conditions to sabotage a nation's harvest. (In 1977, the United Nations recognized the potential for such actions early on, convening 85 countries to sign the "Prohibition of Military or Any Other Hostile Use of Environmental Modification Techniques" treaty). Last, nations can face ethical decisions related to weather modification, such as with the Chernobyl nuclear disaster in the Ukrainian SSR in 1986. Immediately following the disaster, a cloud plume filled with radioactive material was headed towards densely populated Russian cities. Authorities decided to employ cloud-seeding to inject silver iodide into the clouds, eliciting rainfall, which fell onto smaller villages in Belarus instead of Moscow. In other words, "More than 4,000 square miles of Belarus were sacrificed to save the Russian capital from the toxic radioactive material."[19] Nearly 40 years later, the Russian government still denies the cloud-seeding efforts that affected so many Belarusian villagers, yet several Russian pilots have been publicly honored for their roles in diverting the atomic clouds away from the capital city.

Large-scale geoengineering projects will undoubtedly create a power imbalance that protects the privileged and disadvantages more vulnerable populations. A 2021 report published by the U.S. Environmental Protection Agency (EPA) found that those who are considered to be socially vulnerable – based on race and ethnicity, age, income, and education – are significantly more likely to be impacted by the effects of climate change, including changes in air quality, flooding damage, and extreme temperatures, particularly for weather-exposed workers.[20] Carl Sagan, the popular American astronomer, recognized the climate disparity in his 1985 address to the U.S. Congress, in which he said:

> [T]he nations who deal with this problem have to make a change from their traditional concern about themselves and not about the planet and the species... one region of the planet might benefit while other regions of the planet suffer, and there has to be a kind of trading off of benefits and suffering, and that requires a degree of international amity.[21]

As we experiment with weather and climate modification, we must do so ethically. We should be mindful of how our local actions (and non-actions) impact the rest of the world. And there are legitimate arguments to be made that the countries that are historically responsible for climate change and who benefitted most from it also should lead the efforts in climate action.[22] We are not all affected by climate change in the same ways, but as Sagan reminded us, "We are all in this greenhouse together."

Sustainability

We hear the word *sustainability* often. But what does it really mean? In 1987, the United Nations (UN) defined sustainability as "meeting the needs of the present without compromising the ability of future generations to meet their own needs."[23] This definition, of course, is broadly inclusive of all human activity that might impact others. The UN has since developed a list of 17 Sustainable Development Goals (SDG) that it has committed to pursue, including, but not limited to, the general categories of social, gender, and wealth inequality, climate change, environment, peace and justice. Each goal has measurable objectives that it strives to meet.

I see many parallels between each of these sustainability goals and ethical design since both emphasize the mindful well-being of others and prioritize the future over the present. In fact, we can actively respond to many of our present global challenges through the ethical design of sustainable products and habits. *Green technologies*, for example, are those innovations that are designed to mitigate human effects on the environment. We associate green technologies mostly with energy production and conservation; for instance, windmills, solar panel farms, and hydroelectricity plants are green infrastructures that counter our dependence on non-renewable, fossil fuels. Ethical design might manifest itself in large-scale energy projects and initiatives such as these, but it also might be represented in smaller, individual efforts. Consumers do not have to wait around for total energy independence in order to make a difference.

Table 4.1. Comprehensive list of goals

1. No Poverty	9. Industry, Innovation, and Infrastructure
2. Zero Hunger	10. Reduced Inequalities
3. Good Health and Well-being	11. Sustainable Cities and Communities
4. Quality Education	12. Responsible Consumption and Production
5. Gender Equality	13. Climate Action
6. Clean Water and Sanitation	14. Life Below Water
7. Affordable and Clean Energy	15. Life on Land
8. Decent Work and Economic Growth	16. Peace, Justice, and Strong Institutions
	17. Partnerships for the Goals

In fact, it is not realistic to hope for a single green technology to save us from an impending climate disaster. There is no magical solution to mitigating human effects on climate. Rather, we should live more sustainably, by making environmentally conscious decisions in our personal lives. For example, I began this book by using the example of the Keurig K-Cup and its unethical design. Avoiding the single-use plastic K-Cup is one simple decision that consumers can make to help reduce the amount of non-biodegradable waste that goes into the environment. Small lifestyle decisions – e.g., opting for reusable cups and metal straws, recycling and composting, installing high-efficiency fixtures in showers, to name a few – can add up to big change. But even when presented with these smarter, green options, people can be somewhat reluctant to sacrifice personal convenience for the betterment of the environment.

Although studies have shown that most people have the natural inclination to recycle their waste, the failure to do so is actually dependent on a few different variables. First, environmental knowledge and concern are key predictors of recycling behaviors.[24] If consumers are more aware of the environmental effects of non-recycled material, and they are emotionally invested, they also are more likely to engage in positive waste management behaviors. Knowing this, environmental agencies and advocacy groups have used awareness campaigns and public service announcements to educate the public on the benefits of recycling and why it is necessary. Secondly, and perhaps most importantly, recycling opportunities have to be convenient for the consumer. According to the Project for Public Spaces (PPS), the successful use of waste and recycling receptacles depends on their location and design. PPS recommends installing two to four receptacles per city block, though the placement is relative to the average number of people and the types of vendors in the area; receptacles are more useful in high-traffic areas that serve food and drink.[25] Yet another key variable in getting people to recycle is the physical design of waste and recycling receptacles. The images below show how trash and recycling receptacles can be designed effectively to communicate their specific affordances to the user.

Figure 4.1. On the left: Trash and recycling receptacle designs cleverly communicate their intentions. On the right: Trash and recycling receptacles are not clearly distinguishable.

Rather than relying completely on the behaviors of consumers, the image on the left employs a scripted design that requires the user to consciously identify which receptacle to dispose of their waste. The image on the left shows two bins; the yellow recycling bin regulates what types of waste are allowed, inviting items like cans to be deposited through the circular openings, whereas the larger nondescript opening on the green bin beside it can accept general, non-recyclable waste. Of course, the system is not entirely perfect; users still can force smaller, non-recyclable waste through the openings of the recycling bin or mindlessly throw recyclable items into the large-mouth bin. But the design succeeds in being what Martin Heidegger calls "present-to-hand." This concept, you'll remember from the previous chapter, describes how the design of an object can spark a brief epiphany when it does not perform as expected. This momentary pause might be sufficient to encourage the user to deposit their waste in the proper receptacle rather than thoughtlessly throwing their trash into the nearest bin, and this extra nudge might result in collecting more recyclable materials. The image on the right, however, shows waste bins that are nearly identical, with the only difference being a label of which types of waste each bin should accept. This generic design results in waste becoming "contaminated" – the industry term for when non-renewable, foreign material is mixed in with recyclable material. For most municipalities, it is too expensive to sort contaminated waste, and because the amount of contamination has increased significantly over the years, the result is that more than 25% of all recycled materials are diverted to a general landfill instead.[26] When trash and recycling receptacles are mislabeled or appear too similar, confused consumers will simply discard their waste into the bins without sorting properly (i.e., recycling is no longer convenient). Therefore, the National Recreational and Parks Association (NRPA) has determined that "recycling outcomes are aligned to predictable design."[27]

Recycling works when we make deliberate efforts to do so. For instance, in 2019, authorities in Shanghai passed a legal resolution that requires residents to sort their waste. Since the inception of the program, there has been a significant increase in the amount of recyclable material processed and a significant decrease in levels of nitrogen dioxide and carbon monoxide in the air, resulting in a better Air Quality Index (AQI) in Shanghai proper and in surrounding cities.[28] Recycling policies, initiatives, and products are ethically positive, but there might be an even better solution. In 2020, *Politico* ran a special investigative series, "The Recycling Myth," which filed a provocative article titled "How Recycling is Killing the Planet." The report asserts that recycling gives consumers a false sense of environmental hope; their earnest efforts to recycle their waste are honorable but not sufficient in combating climate change. Further, purchasing behaviors are directly influenced by eco-friendly and "green" packaging and labeling, which consumer research has shown to elicit more positive perceptions in buyers. Advertisers and marketers have keenly noted this emotional appeal to the ethical consumer who wishes to minimize their environmental footprint, but many products are misrepresented as being "green" and, in some cases, cause even more environmental harm.

The "direct rebound effect" is the term used by economists and researchers to describe how consumers typically will choose eco-friendly products but use them for longer or at a higher rate, thus increasing their overall "net green." An example is replacing incandescent light bulbs with LED bulbs, only to leave more lights on for longer. The result is a net increase in energy consumption, even though the consumer was led to believe that a more environmentally conscious purchase decision was made. An article in the *Stanford Social Innovation Review* argues simply that, for a variety of reasons across a variety of industries, "There is no such thing as a 'green' product."[29] One study showed that in ten different industrial sectors, energy consumption outpaced efficiency improvements in each area.[30] Instead of being self-satisfied with buying green products, a more ethical move is to redefine our behaviors: fuel-efficient vehicles are not as effective as ride-sharing; processing recycled material is less eco-friendly than abstaining from using these materials altogether; energy-efficient appliances do not grant us license to do the laundry or dishes more often. Perhaps the key to sustainability is not committing more resources to recycling but refocusing our efforts toward waste prevention. And there is no greater example of this than in the fashion industry.

The use of the term "fast fashion" has become, well, fashionable. The European Union (EU) defines fast fashion as "high volumes of lower quality garments at low price levels."[31] An analysis using Google's Ngram viewer reveals that the phrase first appeared in 1998 and continues to rise in popularity. We cannot glean too much from this basic linguistic analysis trick other than that there has been an increased awareness of the "fast fashion" phenomenon over the past two decades, which also happens to coincide with the ubiquity of the internet, online shopping, and mobile devices.[32] The fast fashion industry includes popular manufacturers and retailers like SHEIN, a Chinese company founded in 2012 that consistently ranks in the top five shopping apps for mobile devices and boasts more than 250 million followers on social media. As of November 1, 2023, SHEIN.com is the most frequently visited fashion and apparel website worldwide.[33] SHEIN is a self-described "digital first fashion and lifestyle e-tailer" that offers low-priced clothing and accessories and free global shipping. According to its own Sustainability and Social Impact Report (2021), the company claims to support all 17 Sustainable Development Goals set forth by the United Nations and is even a signatory to the UN Global Compact. But, many have called SHEIN into question for its unethical practices.

SHEIN is notably absent, however, from the United Nation's Fashion Industry Charter, which is a list of fashion and apparel companies that have pledged to reduce their greenhouse gas emissions to net zero by the year 2050. As many have reported, SHEIN emits more than 6 million tons of CO_2 each year – the rough equivalent of 180 coal-burning power plants – and its raw production materials may contain dangerously high levels of lead and other toxic chemicals.[34] The company also has been investigated for violation of labor laws.

Overall, the fast fashion industry is the second largest polluter (behind the oil and gas industry) because of its CO_2 emission levels and the wastewater that it generates. According to the UN, the fast fashion industry is culpable for anywhere between 2 and 8 percent of global carbon emissions; the entire industries of aviation and shipping account for roughly 5 percent.[35] And in some places, such as South Africa, entire rivers run blue from the runoff dye that is used in denim production facilities. But we continue to buy cheap because it suits our needs and our wallets.[36] As with recycling behaviors, it is going to take more than public service announcements to change the purchasing decisions of consumers, and it begins with the manufacturers. The UN has founded the Conscious Fashion and Lifestyle Network, which is a database of stakeholders, companies, and governments committed to sustainable practices in the fashion industry. Similarly, the Conscious Fashion Collective (CFC) spotlights products and brands that are ecologically and socially responsible. And the Environment Programme arm of UN Europe has created the Sustainable Fashion Communication Playbook – a strategic framework aimed at brand managers, marketers, and fashion media – that offers a blueprint for how the fashion industry can attain its goal of net-zero carbon emissions.

Ethical design can play a pivotal role in the climate crisis. We must begin by shining a light on those manufacturers and policymakers who refuse to choose sustainable products over profit. We, as consumers, also must choose ethical products over prices. We can begin by evaluating the things that we purchase, use, and then throw away, starting with the clothing on our backs. By living and participating in the modern world, we are signatories of a social contract – one that pursues a *common good*, for us and for future generations both on this planet and beyond.

Space environmentalism

In 1962, President John F. Kennedy stood in front of a crowd at Rice University and gave one of his most memorable speeches. His "We Choose the Moon" speech was an aspirational monologue that laid out the goal for the United States to be the first country to put a person on the moon by the end of the decade. He said, "We choose to go to the moon in this decade and do the other things, not because they are easy, but because they are hard."[37] Of course, Kennedy was assassinated the following year and was unable to witness the realization of this promise in the 1969 Apollo 11 moon landing. Although the Soviet Union was the first to land an unmanned spacecraft on the surface of the moon ten years earlier, the United States is the only country to have successfully landed a person on the moon. But since Neil Armstrong and Buzz Aldrin first set foot on the lunar surface in 1969, there have only been six manned return missions, with the last taking place in December 1972. By this time, it had become evident that the public had lost interest in space exploration. This, coupled with the economic recession of the early 1970's, effectively grounded all missions to the moon and beyond.

More than fifty years later, NASA has announced that it will be launching another mission – Artemis – to put the first female and person of color on the moon. The mission faces the same criticisms that ended the space shuttle program in the first place: U.S. space exploration is an expensive endeavor funded by its taxpayers, who see little benefit from moon landings. NASA's inspector general has acknowledged that the project is already over budget, estimating that the Artemis program will cost taxpayers nearly $100 billion over a period of 13 years, with a single launch costing over $4 billion.[38] This extravagant spending has some pointing out that we could do a whole lot of good here on Earth with that kind of budget. Even still, NASA asserts that there are three distinct benefits to returning to the moon: "scientific discovery, economic benefits, and inspiration for a new generation of explorers."[39] In fact, many see the Artemis moon missions as a stepping-stone to putting the first person on Mars.

Neil deGrasse Tyson is a widely known astrophysicist and has been called the "most captivating scientific communicator."[40] Tyson has long been a vocal advocate for the U.S. to resume its space exploration, despite the enormous price tag. In 2012, he published an article titled "The Case for Space: Why We Should Keep Reaching for the Stars," in which he articulates the many reasons why space exploration is economically, militarily, and philosophically feasible. A central plank in his platform is that the pursuit of space exploration yields "incalculable" benefits. These gains are both intangible, such as in intellectual stimulation and curiosity in younger generations, and tangible, such as in the development of new technologies (e.g., GPS, LASIK eye surgery, digital imaging, and pacemakers, to name a few).[41] Mike DiCicco of NASA's Goddard Space Flight Center writes, "Technology created for Artemis will certainly find secondary applications on Earth."[42] Still, people have reservations about returning to the moon. On a recent episode of a popular podcast series, Universe Lair, Tyson was asked what U.S. taxpayers might gain by going back to the moon. His response was as follows:

> When we went to the moon, to explore the moon and turned around and photographed Earth, we discovered Earth for the first time. We went to the moon to explore the moon and we discovered Earth... as the universe intended you to see it. And we're all down there together. Only when that photo of Earth from the moon was published did we say 'Maybe we should ban leaded gas. Maybe we should ban DDT. Maybe we should clean up the Earth. Maybe we should clean up the air.'[43]

Tyson goes on to point out that environmental legislation was reformed, and new agencies were established as a result of our trip to the moon and the subsequent realization that we need to take care of our planet. The Apollo missions gave us more than moon rocks; they gave us perspective.

During this same period of space exploration, a major oil spill off of the coast of Santa Barbara, California, and other ecological disasters prompted U.S. President Richard Nixon to establish the Environmental Protection Agency (EPA), whose central mission is to "protect human health and the environment."[44] The bipartisan-supported agency touts many accomplishments including prohibiting the sale of lead-based paint, banning harmful pesticides like DDT, establishing carbon emission standards, and passing comprehensive environmental legislation such as the Clean Air Act (1970), the Ocean Dumping Act (1972), Safe Drinking Water Act (1974), Clean Water Act (1977), Nuclear Waste Policy Act (1982), Pollution Prevention Act (1990), Food Quality Protection Act (1996), among others. The EPA, in partnership with other international environmental agencies such as the European Environment

Agency (EEA), Global Alliance on Health and Pollution (GAHP), and United Nations Environment Programme (UNEP) and in accordance with international environmental agreements, such as the Paris Climate Accords and the Kyoto Protocol, seeks to preserve human life and the natural environment here on Earth.

Space environmentalism, then, argues that our natural environment should be extended to include orbital space (defined as the airspace between 100km and 36,000km above Earth's surface) and that we make a concerted effort to protect it just as do with terra firma. A 2022 article published in the journal *Nature Astronomy* and titled "The Case for Space Environmentalism" features a cohort of astronomers, researchers, professors, and astrophysicists who meticulously and credibly outline the threats posed by an ungoverned and unregulated orbital space, likening it to the catastrophic environmental negligence that went unchecked on Earth for most of human history. As of 2018, there were roughly 2,000 active satellites in Earth's gravitational orbit; since then, Elon Musk's company, SpaceX, has single-handedly doubled that number. The result? More objects jockeying for position in an already crowded layer of space. In fact, the General Catalog of Artificial Space Objects (GCAT) reports more than 23,000 objects (the size of a softball or larger) in low-Earth orbit (LEO).[45] GCAT's database goes back to 1958 and shows that the number of objects in orbit continues to grow exponentially year after year. The colloquial term for this orbiting debris field is "space junk," and it can cause major problems. For one, space junk can collide with critical communications satellites and spacecraft. The European Space Agency reports that this happens about once every ten years. The second threat posed by space junk is that it will eventually return to Earth, potentially killing someone (as the FAA predicts will happen once every two years on average, by the year 2035)[46] or re-introducing harmful contaminants and metals into our atmosphere (as theorized in a 2023 study published in *Earth, Atmospheric, and Planetary Sciences*[47]). For these reasons and many others, conscious space environmentalism is necessary. Just as we strive to protect our planet, we also should take a proactive rather than reactive approach towards a sustainable and ethical future beyond the globe.

Ethical design frameworks ought to be applied to space environmentalism by evaluating current and future space technologies for their impact and ethical positivity. For example, although Musk's SpaceX has contributed a significant number of satellites to low-Earth orbit (an already overcrowded debris field), these satellites also provide crucial support to underserved and vulnerable regions of the globe. The company's subsidiary, Starlink, has supported

Ukrainian forces and humanitarian aid groups in the Gaza Strip with essential communications and internet connectivity during times of war. However, questions remain about the ethical nature of a privately owned for-profit company such as SpaceX and its eccentric, and at times erratic, owner making sensitive geopolitical decisions without formal diplomacy efforts. For instance, Musk has acknowledged that he denied Ukraine's request for access to the Starlink network so that their military could execute a surprise attack on Russian forces stationed in the Crimean Peninsula, fearing that he would be "explicitly complicit in a major act of war and conflict escalation."[48] There is a new age space race happening now, but instead of entire countries competing for galactic dominance, it is between the proprietary satellite programs of Elon Musk (Starlink) and Jeff Bezos (Project Kuiper). And what are we to do when a powerful billionaire's personal ethics, values, or morals clash with the common good?

Figure 4.2. A photo of space junk in low-Earth orbit (LEO).

A strong argument can be made that internet connectivity is becoming an essential human right – on par with our basic needs for food, water, and shelter. Underdeveloped regions and populations are especially vulnerable to this need, as they often depend on the internet as a conduit to accessing basic human rights such as employment, education, housing, and other social services.[49] Back in Chapter 1, I introduced the Ethical Hierarchy of Needs, which is built upon the foundation of Human Rights. Without this building block, the ethical well-being of individuals cannot be achieved. Just as we might give pause to a system where a single person or entity governs our food or water supply, we might also question whether it is ethically permissible to do so for internet access.

Aside from the doomsday scenario of inciting world war or governing internet connectivity for entire regions of the globe, commercial satellite programs pose other ethical threats. There is the obvious problem of over-polluting the low-Earth orbit, which increases the risk of collision with space junk and endangers space travel for astronauts, but there also is the problem of *visual pollution* – the aesthetic quality of the observed landscape, both natural and man-made. The more objects we launch into low-Earth orbit, the more obfuscated the night sky eventually becomes. This poses scientific challenges for astronomers and observatories here on Earth, but it also introduces new questions about our individual right to an unobstructed sky view.

We designate areas such as National Parks both for their ecological conservation as well as for our personal wonderment and enjoyment. Millions of people visit these protected environments because of their natural beauty and the intrinsic value that they hold. In the face of sprawling commercialization, we insist on protecting these environments, just as nature intended. Perhaps there is no greater example of this than Central Park in New York City, a natural area of roughly 843 acres (3.41 km^2) located in the heart of one of the most densely populated cities with the most expensive real estate in the world. Just as the Central Park Conservancy group is dedicated to the preservation of the green space for the "enjoyment and wellbeing of all,"[50] the group DarkSky International works to sustain the nighttime environment sky through legislative policy, scientific research, and public advocacy campaigns. The rationale is that the sky above should be preserved for all to enjoy and is not a blank slate for commercial satellite programs to occupy.

A Russian startup company, StartRocket, has proposed launching thousands of tiny satellites called CubeSats into low-Earth orbit. These CubeSats would be controlled remotely to be organized like pixels on a screen, forming massive billboards in the sky. PepsiCo Russia was one of the first advertisers to experiment with StartRocket; a spokesperson for the company said, "Orbital billboards are the revolution on the market of communications."[51] But aside from the annoyance of visual pollution, some have raised other ethical concerns about orbital marketing. For example, the nighttime sky has long held cultural value and significance since the beginning of mankind, playing an essential role in storytelling and ancestry. Treading on the sky with orbital advertisements is analogous to planting a giant, flashing billboard for Pepsi on an Indigenous people's sacred land. We have an ethical obligation to our natural environment, and this includes space.

Case study: Wave pools

The life of a surfer on the east coast of the United States can be rough. I should know because I am one. Having surfed for over 20 years, I can tell you that the surf in North Carolina is fickle; it can be as flat as a lake on hot, summer days, but the continental shelf and shallow beach breaks of the Cape Fear also can produce immaculate, overhead waves on occasion. Hurricane season generates the gnarliest and most consistent waves, and in the off-season, East Coast surfers find themselves traveling all over the world in search of surf. Our passports are filled with stamps from trips to remote areas in Central American countries, and the most dedicated are prepared to hop on quick flights to Puerto Rico to swell-chase. My children, who also surf, often complain when there is no rideable wave down the street from our house. I explained to them that surfing is not like other sports; we don't have a field, a gym, or a court that we can visit at any time to quench our desire to surf. We surf when there is surf, not just whenever we feel like it. Surfers constantly monitor weather, wind, swell patterns, tide cycles, and other meteorological conditions that dictate whether there is good surf. We are at the mercy of nature, and so we look for creative substitutions. Skateboarding, kitesurfing, hydrofoiling, and wakeboarding were all borne of desperate surfers looking to manufacture the thrill of wave riding. Others have even discovered more extreme activities like glacier surfing (this is where a surfer waits in arctic waters until a giant ice shelf breaks off a glacier and creates a rideable wave), river surfing (in certain parts of the world, rivers can be surfed when they experience flash floods or have extreme tidal bores) and ferry surfing (chasing down large ferry boats to ride their wakes). In addition to searching for novel waves, we are developing more sophisticated ways to mimic them.

There is a storied history of people designing technologies to generate manmade waves. The first known attempt was in Bavaria in the 1870s. King Ludwig II enjoyed the relaxing sound of rippling waves in his personal grotto, so he built a "wave maker"[52] that generated tiny disturbances in the water. Far from surfable, this marked the start of an iterative process to replicate the ocean's waves. Throughout the twentieth century, wave pool technology progressed in places like Germany, England, Japan, and the United States, using a variety of hydraulic pumping systems that could move voluminous amounts of water that simulated waves. Moving higher volumes of water, more quickly, would produce a larger wave. By the end of the century, artificial wave pool technology had been refined to create a surfable wave that allowed surfers to ride a roughly 3–4-foot wave for up to 20 seconds. Today, wave specifications like size, length of time, and shape are possible and can be changed within a matter of minutes to suit the needs of the surfer, and artificial wave pools can be found all over, even on cruise ships. Wavepoolmag.com has developed a comprehensive interactive map that shows more than 200 completed, in-progress, and proposed artificial wave pools throughout the entire world. Wave pools can be found in unlikely landlocked areas like Edmonton, Canada; Waco, Texas; Sun City, South Africa; and Anyang, China. But they also can be surfed in locations that already have epic natural oceanic surf, such as in coastal Japan, Spain, and even the Hawaiian Islands. A wave pool can be positioned against a desert landscape (as in Abu Dhabi, UAE) or inside a shopping mega-mall (as in East Rutherford, NJ, USA). Indeed, an artificial wave pool can be found on every continent in the world, minus Antarctica. Kelley Slater, widely regarded as the greatest living surfer, has founded the Kelley Slater Wave Co., which has partnered with the World Surfing League (the professional surfing tour) to host professional surfing events at its flagship location, the Surf Ranch in Lemoore, California. It is the only artificial wave pool on the circuit.

There are many variations of wave pool designs.[53] All require a giant body of water that is agitated by some external mechanism. For instance, some wave pools utilize compressed air chambers that forcefully push water out and over a sculpted bottom. It is this sudden change in bathymetry – the floor of the body of water –that creates the wave. Other designs, like the Surf Ranch, drag a hydrofoil underneath the surface, displacing water and causing a swell that can be ridden. Surf Lakes, located in Queensland, Australia, drops a giant plunger in the center of a lake, creating a fast-moving swell of water that breaks along synthetic reefs that are designed to produce specific types of waves for different skill levels. Still more innovative designs implement underwater rotating fan blades and water dumping to simulate a perfectly shaped wave.

Of course, all wave pools, regardless of their functional design, necessitate energy. Each wave pool draws different amounts of energy, though some designs are more energy efficient than others. Wavegarden claims to be "the most energy-efficient wave generation technology" but an analysis revealed that the Wavegarden wave-generating technology uses 300 to 400 kWh per one-hour surf session.[54] For reference, the average American uses approximately 10,660 kWh each year, which translates to roughly 1.2 kWh per hour. Wavegarden technology

– the self-anointed "most energy-efficient" – uses up to 400% more energy than a typical energy consumer in the U.S. This can seem overindulgent and potentially irresponsible, especially in cases like the BSR Surf Ranch in Waco, Texas, which relies on an infamously unreliable power grid that has a history of failing millions of people in severe weather, such as in 2021 when a blackout resulted in at least 57 deaths and nearly $200 billion in damages.[55] Further, wave pools require water, which is an increasingly scarce resource in places. Water scarcity is relative to specific locations, but it disproportionately affects underserved and vulnerable populations, with poorer communities impacted most significantly.

Advocates for artificial wave pools point out that the technology can bring surfing inland to disadvantaged areas and help diversify surfing, an activity usually reserved for those who live on the coast and have access to beaches. This may be so, but surfing organizations like Surfrider Foundation Europe, a nonprofit company created by and for surfers, have taken an official position against artificial wave pools because their "environmental concerns outweigh their value."[56] Surfers themselves are conflicted about their support for wave pools for various reasons; on the one hand, wave pools guarantee rideable surf, but on the other hand, do they constitute authentic surfing? To better understand the ethical design of wave pools, I interviewed Dr. Peter Singer, Professor of Bioethics at the University Center for Human Values at Princeton University, for his opinion on the matter. Singer has been called the "planet's most influential philosopher"[57] and has written numerous books and articles on ethics. The 77-year-old Australian is also an avid surfer. Singer is not ambivalent towards artificial wave pools, telling me that he has "no interest in surfing them" because they detract from the intangibles that ocean surfing offers. Wave pools do not offer the same sweeping vistas, the unpredictability of waves, and the natural energy of the ocean that courses through a surfer's veins. For these reasons, Singer finds it difficult to endorse the man-made wave. However, he does recognize that this building trend is growing in popularity and that, eventually, opportunistic capitalists will find a way to monetize waves. If we are going to build artificial wave pools, Singer argues, we must ensure that they "use 100% green energy and are in places where there is abundant water, and they are replacing, say, obsolete factory sites rather than natural habitat" (Peter Singer, personal communication, November 25, 2023). Only under these circumstances can artificial wave pools be ethically defensible.

As wave pools become more ubiquitous in landlocked areas, we must consider this ethical posture. As a surfer, I find myself torn; the allure of a perfectly shaped wave that can be conjured up on demand is worth whatever the price tag. But the guilt quickly sets in. The opulence of an energy-sucking wave factory seems irresponsible when so much of the world's population lacks basic access to potable water. Surfers often talk about their harmonious relationship with nature; when done properly, surfing leaves no environmental footprint. What makes surfing unlike other sports is that it seizes on a fleeting moment in a specific place and time when a natural wave appears suddenly on the horizon, is ridden briefly by a surfer who is well-positioned, and then dissipates into shore break, as if it never happened. The process is infinitely sustainable under the right conditions. For centuries, surfing has been more than catching a perfectly simulated wave; it is about the beautiful symbiosis between humans and their natural world. There is no doubt that engineers and builders will try to imitate this natural phenomenon through design and construction, but we must ask ourselves whether it is worth the cost or if some things are better left unrivaled.

Chapter 5

Crime

The physical environment can send a powerful signal about who, and what, we are.[1]
Emily Anthes

Suppose you are walking down the sidewalk and spot a $5 bill on the ground. Do you pick it up? Do you then scan your surroundings for its rightful owner, or do you pocket it and keep on walking, pleased with your good fortune? How about a $100 bill? Do you pick it up, and if so, are you just as quick to stash the money in your pocket, or do you put a little more effort into trying to return the money to its owner? What exactly is the ethically responsible behavior in this situation? It turns out it depends. In 2019, researchers at the University of Michigan conducted a study of human honesty across 355 major cities in 40 countries. The premise of the experiment was basic: Research assistants would pose as someone who had just found a lost wallet in a public place and then approach an unsuspecting person, saying, "Somebody must have lost it. I'm in a hurry and have to go. Can you please take care of it?" The wallet contained various items, including business cards with contact information and various amounts of money ranging from $13.45 to $94.15.[2] Then, the researchers waited to see which people would attempt to contact the owner of the wallet and return it. Their null hypothesis (assumption) was that the more money in the wallet, the more likely the "finders" were to keep it for themselves. But the results were exactly the opposite; the wallets with more money in them were *more likely* to be returned than those with less money. The researchers concluded that "psychological forces —an aversion to not viewing oneself as a thief— can be stronger than the financial ones."[3]

Ultimately, behavioral ethics comes down to our own self-perception of right and wrong. Dan Ariely, cognitive science professor at Duke University and author of *The (Honest) Truth About Dishonesty* (2012), writes: "Essentially, we cheat up to the level that allows us to retain our self-image as reasonably honest individuals."[4] That is, we'll permit ourselves to behave relative to our own moral thresholds, rather than naturally adhere to some universally ethical code, as long as we consider our actions justifiable. This, by the way, reinforces Thomas

Hobbes's view of humanity's default "state of nature" and the subsequent need for a social contract. One crucial assumption that the experimental research study makes is that the keepers of the "lost wallet" kept the money for themselves. But what if I tweak the scenario just a bit: What if the finder is handed a wallet with $100 in it and told to "take care of it," but instead of returning it to the owner, they donate the money to charity? Does this justify the original behavior? This question is purely utilitarian in that it accounts for the *intention* of the wallet-finder rather than the isolated act of finding and keeping the wallet, regardless of how the money is used.

One could even make the argument that the circumstances surrounding the found wallet scenario could influence the finder's behavior. The situational context of the experimental study was that the participants were forced into deciding. They were approached directly and handed a "lost wallet"; they did not actively pursue a stray wallet lying on the ground. By being handed the wallet, the participants were required to make a binary decision to either keep it or return it, but a third option would have been to ignore the wallet altogether (though you also could make the argument here that inaction would, in fact, be a decisive action). Just as the affordances and the scripting of objects can suggest behaviors to users, so too can the situational context of their surrounding environment. For example, suppose you spotted a wallet or a $100 bill lying on the ground, but it was in the middle of a busy street and retrieving it would be risky: Are you more or less likely to pick it up? Or suppose you are walking down the street, and an armored Brinks truck is involved in an unfortunate accident with another vehicle (thankfully, no one was hurt), causing its rear doors to fling open, spewing hundreds of $100 bills into the air, landing right at your feet. Wouldn't you be more likely to pick up this money? And if you knew, for instance, that the money had come from a nearby bank that is FDIC insured, rather than from the purse of a single mother who had dropped the money on the ground accidentally, wouldn't this also factor into your decision-making?

Filmmakers often exploit these dubious ethical choices in bank heist movies, where viewers find themselves sympathizing with and rooting for the robbers, somehow justifying their illegal behavior as some form of warranted rebellion against the establishment. In the 1990 crime thriller, *Point Break*, an FBI agent (Keanu Reeves) becomes enthralled with a gang of surfer bank robbers, whose leader, Bodhi (Patrick Swayze), gives an impassioned speech to his crew:

> This was never about the money; this was about us against the system. That system that kills the human spirit. We stand for something. To those dead souls, inching along the freeways in their metal coffins, we show them that the human spirit is still alive.[5]

By the end of the film, the straight-arrow FBI agent has left the Bureau and adopted Bodhi's enlightened way of life (minus all the robbery, kidnapping, and killing). Similarly, in Michael Mann's film *Heat* (1995), Robert DeNiro's character walks into a bank, assaults several customers and security guards, and then announces, "We want to hurt no one. We're here for the bank's money, not your money. Your money is insured by the federal government; you're not gonna lose a dime." Here, armed bank robbery is presented as a victimless crime. Countless other films in the heist genre frame robbers as protagonists. The casts of the Ocean's movies (*Ocean's Eleven, Twelve, Thirteen,* and *8*) endear audiences into rooting for their successful capers. Art heist films like *The Thomas Crown Affair* and *Entrapment* present the art thief as a debonair Robin Hood-type figure. A 2023 film, *The Naughty Nine,* is a family comedy about nine children forming an adorable scheme to rob Santa's workshop in the North Pole. Another film, *The Monuments Men,* tells the true story of a group of WWII Allied soldiers as they confiscate priceless, culturally significant objects before they can be stolen or destroyed by the Nazi regime. This is all to say that grand theft can be ethically and morally justifiable under the right conditions.

So, you see, we must account for the circumstances (and designs) that impact behaviors – even if criminal in nature. Theft – and more serious crimes like killing – are not wrong in absolution; they become ethically justifiable at times. Just as the Trolley Problem from Chapter 2 illustrates, the context in which these actions take place determines the morality of our choices. Understanding this, we might consider whether our built environments – buildings, complexes, outdoor spaces, and even infrastructure – encourage criminality, presume guilt, and punish unreasonably. And if so, could the inverse be true? Can we instead design our built environments for ethically positive decision-making and less crime?

Designs against crime

Robbery is a design issue. Or at least this is what Geoff Manaugh argues. Manaugh is a bestselling author and curator of the architecture website BLDGBLOG and pays particular attention to the relationships between technology, infrastructure, and design. His 2016 book, *A Burglar's Guide to the*

City, investigates the seedy underworld of burglars, thieves, and robbers, as seen through the lens of design and architecture. Manaugh carefully profiles the painstaking methods of notorious robbers, some of whom have pulled off some of the greatest heists in history, before ultimately arriving at the book's conclusion: "Burglars use cities better." Putting aside for a moment the ethical principle that theft is wrong, Manaugh draws a straight line between *burglary* and *design* – specifically, the physical design of architecture and infrastructure. In fact, without architecture, burglary would not even be possible. For this statement to be true, we must understand what burglary is. The FBI gives an extremely vague definition of burglary as committing a theft via unlawful and forceful entry into a "defined architectural space" – even if that space is air. For instance, burglary can be forced entry into a locked jewelry store, but it also can be the act of taking an item from a parked convertible with its top-down or from an open-air vegetable stand at the farmer's market (this action, in legal terms is referred to as "breaking the close," where the "close" is a physical area that defines a space that a burglar enters illegally). Essentially, burglary is a "spatial crime" that takes advantage of *closes* that are embedded into architecture.

Burglars are interpreters of the built environment. Each of the aforementioned *Ocean's* heist movies (*Ocean's Eleven, Twelve, Thirteen,* and *8*) begins with an inciting event that necessitates a seemingly impossible theft that involves Las Vegas gambling casinos. Facing an assortment of twists and turns, setbacks, and double-crosses, the crew always manages to find ways to overcome the heavily layered and high-tech security measures of these buildings, suggesting that no building can be entirely secure; burglary is just an elaborate puzzle waiting for criminals to solve. Perhaps this is why Manaugh describes burglars as "rogue usability experts" who are just "idiot masters of the built environment." Burglary requires an acute awareness and deep understanding of the built environment. Every good heist movie features a scene or montage showing thieves pouring over blueprints of buildings, making scaled models, and surveilling locations (also known as "casing the joint"). In both the movies and in real life, complicated heists are meticulously planned, not spontaneous, because they must outsmart designs meant to prevent theft.

Good burglars, according to Manaugh, figure out ways to exploit the built environment that surrounds them. He writes, "The city gives off signals – provided you know how to read them."[6] I've already detailed, in Chapter 3, how architecture deliberately communicates certain affordances to us: Proceed this way, use this door, wait here, etc. Well-designed public spaces like airports, parks, and retail stores demand specific behaviors from their users, compelling them to act in specific ways. We can examine the ethics behind these designs, and we can debate whether it is unethical for a retail store or a supermarket to manipulate shoppers' behaviors to increase their bottom line. Similarly, we might question whether it is ethical to entice people into criminal behavior.

One such controversial example is law enforcement's use of "bait cars." A bait car is a vehicle that has been strategically positioned in a high crime area and left unlocked with the keys inside. It is equipped with GPS-location tracking, cameras, and remote operability that allows officers to disable the car and lock the doors, effectively trapping the would-be car thief. The polarizing tactic yields mixed outcomes and, in some cases, has resulted in officer shoot-outs[7] and pedestrian deaths.[8] The use of bait car programs continues in many countries around the world, though the ethical implications are obvious: Isn't tempting criminals with stealing an unlocked car the same as police *entrapment*? This is both a legal and a philosophical question. Surprisingly, the legal answer is no: "There is no entrapment where a person is ready and willing to break the law, and the government agents merely provide what appears to be a favorable opportunity."[9] But the philosophical answer (as I argue throughout this book) is that, yes, at some point, the design does surpass being a "favorable opportunity" to the would-be criminal and places someone into an unreasonable situation that unduly compels them into illegal behavior.

At the start of this chapter, I asked what you would do if a Brinks truck exploded nearby and $100 bills rained down from the sky, landing at your feet, and no one was there to see you pick up the money. Some model citizens might keep on walking, but others might argue that it would be stupid *not to* pick up the money, even though it constitutes theft. Manaugh writes in his book, "The design of a city might inspire – perhaps even require – certain crimes." To be clear, I'm not advocating for criminal behavior, nor am I blaming design as a scapegoat for bad actions; I'm arguing against weaponizing the built environment to *create* criminals rather than *catch* them.[10]

Inarguably, we can create conditions under which crime is likely. Or, to put it another way, we can design for crime. And so, the opposite must be true. The design also can be used to thwart burglary and other crimes. Manaugh's book gives examples of how architecture can be leveraged to prevent crime. He says, "Whole classes of attack can be ruled out by architectural design alone."[11] For example, a common approach used by criminals is to gain access to their target, such as a bank vault, through an adjoining wall or room. A counter-design measure, Manaugh describes, is a design where the vault is mounted to the exterior of the building, high in the air and in plain sight. This eliminates the ability to gain access from any side but within the room and disincentivizes criminals because of its high profile. In order to deter crime, we must think like criminals.

Crime Prevention through Environmental Design (CPTED) is a practice that emphasizes planned designs for private and public spaces to minimize the likelihood of crime. According to the International CPTED Association, CPTED strategies "reduce victimization, deter offender decisions that precede criminal acts, and build a sense of community among inhabitants so they can gain territorial control of areas."[12] The association has worked closely with the European Committee for Standardization (CEN) to develop standards for residential and urban environments to help safeguard against various crimes. The CEN includes 34 member countries and more than 200,000 technical experts who have authored guidelines on how to design against crimes involving private spaces like personal dwellings and public institutions like schools, retail and office buildings, public transportation, and petrol stations. The standards continue to be developed and offer counsel on preventing specific types of attacks, like the criminal use of vehicles for ramming into buildings and storefronts; this is known as hostile vehicle mitigation (HVM) and includes the strategic placement of vehicle security barriers (VSB) and anti-ram barriers that range from simple stanchions ("bollards") to more advanced systems like wedge barriers that are installed flush with the surface road and hydraulically raise up to form a perimeter around a secure building or area.

Figure 5.1. The Quadrant on Regent Street in London, England, exemplifies CPTED.

We don't necessarily have to fortify all our spaces with criminal deterrents such as gates, fences, spiked roadways, and barriers. I suppose if our only goal was to achieve safety from crime, then all our homes and public buildings would be constructed like prisons – mostly windowless, heavily surveilled, and comprised of unwelcoming and impregnable industrial materials like concrete and steel. No, the artistry of CPTED lies in minimizing crime through subtle design efforts. In fact, it is very likely that you live in an area that utilizes CPTED strategies without you even realizing it. One rather unnoticeable design strategy is the use of curved city streets, as displayed above.

The winding path of Regent Street was plotted in the early 1800s under the direction of Prince Regent, but this was not exactly an intentional design to prevent crime. In fact, the road was meant to connect two popular locales at the time – Carlton House and Regent's Park – in a straight line, but there were land ownership issues, so town planners had to meander around parcels of private property. The result was a serpentine route lined with private residents and high-end retailers whose street-facing windows can feel like constant surveillance. As a result, Regent Street presents a high level of difficulty to burglars hoping to flee from a scene quickly and unnoticed. The curved street requires that a driver proceed cautiously while peering around the corner, and the hundreds of windows peering onto the street might as well be eyewitnesses. Contrast this with a network of straight roadways that allow for easy off-and-on access and traveling at high speeds, as is the case in urban Los Angeles and its expansive freeway system, which invites an average of roughly 1,000 car chases each year by police.[13] The urban grid of a city reflects the types of crimes that occur there.

In fact, research into the correlation between neighborhood street designs and crime suggests a strong relationship between the two. A 1977 report from the U.S. Department of Justice found lower rates of burglary on streets that were less accessible to main roads. Neighborhoods with cul-de-sacs, dead-ends, and L-type street layouts were less likely to experience burglary compared to neighborhoods with "highly permeable" street layouts in which a burglar could easily access and then retreat from.[14] Studies on crime consistently reveal that home burglaries correspond to the number of streets leading into the neighborhood in which they reside; the greater the access from tributary streets, the greater the likelihood of crime.[15] If burglars choose the path of least resistance, then one design against crime has been the gated community.

It is estimated that in 1997, there were around 20,000 communities in the United States that were fortified by gated entrances (roughly 3 million homes, or about 9 million people[16]). Today, there are more than 11 million U.S. homes located in gated communities – representing a 267% increase over the past 25 years.[17] And this phenomenon is not unique to the United States; gated communities continue to experience a similar meteoric rise globally. The rationale for the gated community is predicated on *safety*. But we might question whether this design trend has unintended ethical consequences. The authors of *Fortress America: Gated Communities in the United States* write that this "fortress mentality" does more than just protect its inhabitants:

> Gated communities manifest a number of tensions: between exclusionary aspirations rooted in fear and protection of privilege and the values of civic responsibility; between the trend toward privatization of public services and the ideals of the public good and general welfare; and between the need for personal and community control of the environment and the dangers of making outsiders of fellow citizens.[18]

The authors question whether the rise in the number of gated communities isolates, polarizes, and excludes us from one another, threatening the very fabric of our social contract. While gated communities offer another layer of security from outsiders, they also create "othering," increased inequality, and culture and class-based segregation. An opinion piece published in *The New York Times* even cited the "Gated Community Mentality" as a contributing factor to the killing of Trayvon Martin, a 17-year-old Black male who was fatally shot after a neighborhood watchman identified him as a suspicious person outside of a gated community in Florida.[19] When we design against crime, we must do so in ways that deter criminals without compromising ethical positivity.

Theft and robbery are often motivated by *opportunity*. Ethical designs counter this motivation inconspicuously and without making us feel like we are constantly in danger. One such example of an ethical design against crime is the handbag clip. You may have noticed these in use at restaurants, bars, and cafes; rather than slinging a purse or handbag over the back of a chair, a customer hangs their valuables from the clip, which is in front of them mounted on the underside of the table or on a wall in plain view. There are several different brands that have variations in their design, but many are spring-loaded, requiring the use of two hands to remove the personal item from the enclosure. This simple yet effective design merely relocates the valuable item and makes it a bit more difficult for a passerby to snatch it. The use of the clip has been shown to decrease the number of petty handbag larcenies by half, and it does so without compromising the well-being of others.[20] This is important because we must safeguard ourselves from crime, but we must not treat everyone as a potential criminal.

Pre-criminality

American author Philip K. Dick published his short story, "The Minority Report," in the popular science fiction publication *Fantastic Universe* in 1956. The story was later adapted into a successful film by the same name in 2002, in which the main character, John Allison Anderton, was played by Tom Cruise. Chief Anderton heads the Precrime Division of a law enforcement agency that uses "pre-cogs" (short for *precognition* or knowing beforehand) to predict crimes before they occur. In doing so, crime is virtually eradicated because it is pre-empted by law enforcement. Setting aside the many problematic timeline issues, plot holes, and logical fallacies that plague this story, the method of apprehending pre-criminals poses an interesting paradoxical and metaphysical dilemma – one that Dick acknowledges in this line from Chief Anderton:

> You've probably grasped the basic legalistic drawback to precrime methodology. We're taking in individuals who have broken no law... We claim they're culpable. They, on the other hand, eternally claim they're innocent. And, in a sense, they are innocent.[21]

Precrime relies on a conclusion that is yet to happen and presumably predetermined. This is the definition of a well-known fallacy, c*ircular reasoning*. Chief Anderton seems to spot the logical contradiction, adding that "in our society, we have no major crimes, but we do have a detention camp full of would-be criminals."

Dick's science fiction story isn't that far-fetched after all. In England, it has been reported that at least 14 policing divisions have adopted software that utilizes "predictive mapping" and "individual risk assessment" algorithms to identify and predict the likelihood of crime in specific areas and by specific individuals.[22] The program stops short of arresting or apprehending individuals before committing a crime (as in the Precrime world of "The Minority Report"), but skeptics argue that using computer-driven programs for predictive policing is blatantly unethical. Using our Checklist for Ethical

Evaluation of Systems, Products, and Processes (found in Chapter 2), we can see that predictive policing software fails to meet the ethical design standards in several areas, including a lack of transparency, accountability, and privacy, as well as a failure to account for the values and wellbeing of individuals. Further, predictive policing software is subject to *automation bias*, where the human user unquestionably accepts the correctness of the technology outputs.

Much has been written about the ethical quandaries of predictive policing. Companies like PredPol, the "market leader in predictive policing" that patrols one out of every 33 people in the United States, assert that their tactics are safer and more cost-effective than traditional policing. But using data analytics to predict where crime is likely to occur and by whom seems very similar to *profiling* – a method of predicting behavior based on characteristics and behaviors of a person. Importantly, the law distinguishes between *criminal profiling* and *racial profiling*, where the former is when law enforcement pursues individuals based on "indicators suggesting that someone may be engaged in criminal activity,"[23] and the latter is when law enforcement "targets individuals for suspicion of crime based on the individual's race, ethnicity, religion or national origin."[24] Criminal profiling is legal; racial profiling is not. And if you're wondering how we tell the difference whether someone has been criminally profiled or racially profiled, you're not alone. In the U.S., BIPOC (black, indigenous, and other people of color) are more likely than Whites to report that they have been unlawfully targeted and stopped by law enforcement.[25] This suspicion is warranted: Decades of data on law enforcement programs such as "stop and frisk" in New York City show how BIPOCs are disproportionately affected by law enforcement. The majority of those who were randomly stopped are BIPOC, and a vastly disproportionate number are Black, despite their underrepresentation in the population.[26] Stop-and-frisk activity was also implemented at higher rates in BIPOC and poorer communities.[27] The data would suggest that the randomness of these types of policing programs isn't so random as much as it is strategic racial profiling.

Predictive policing software promises to correct this human bias with machine learning and algorithms, but these, too, are subject to an unethical design. For example, one of the predictors used in the software is arrest data.[28] Communities with higher arrest rates are more likely to be recommended for more policing, and more police leads to more arrests.[29] Just as in Dick's "The Minority Report," this digital redlining is circular reasoning; these data fail to account for whether the original arrest data was biased. If we are inputting bad data to make predictions, the output is more bad data. In other words: Trash in, trash out. Along these same lines, predictive policing relies on unproven technologies like facial recognition software to identify potential criminals based on their appearance and behavior. In Japan, for instance, the National Police Agency has adopted artificially intelligent cameras that scan their surroundings for people who are suspected of carrying weapons and who exhibit any "suspicious" or "unusual" behaviors.[30] The ethical implication here is that AI-assisted technology can easily mistake harmless objects for weapons and misinterpret behaviors as unusual, thus instigating a police intervention on law-abiding citizens. Consider, for example, how someone carrying an oxygen tank for medical purposes or someone shaking involuntarily and spastically from Parkinson's disease might be wrongfully identified as exhibiting "unusual" behavior. Predictive policing can quickly become discriminatory. Facial recognition software has been shown repeatedly to be inaccurate at differentiating between Black faces.[31] Yet, the Police Department in Detroit, Michigan, continues to engage in "dragnet surveillance" using faulty facial recognition software that has resulted in three wrongful arrests in as many years.[32] A major flaw in the predictive policing model is that it assumes that the solution for reducing crime is *more policing* rather than more ethical approaches.

Recently, the European Parliament adopted the Artificial Intelligence (AI) Act, which bans criminal prediction systems and predictive policing using AI technology. Specifically, the Act prohibits law enforcement from using a "person's location, or past criminal behaviour"[33] to predict future behaviors. This legislation also outlaws "remote biometric identification," or facial recognition technology, in public spaces. The AI Act is the first of its kind. However, the predictive policing market is projected to grow substantially in the coming years, as artificial intelligence systems and facial recognition software become more accessible and affordable, despite their obvious problems. One report found that police already spend more time engaged in predictive (or proactive) policing than they do responding to calls for help.[34] But if everyone is assumed to be a potential (or pre-) criminal, then we are already living in Philip K. Dick's sci-fi world of Precrime and punishment. If we are constantly under surveillance, then we forfeit our basic human right to privacy and our presumption of innocence. Even the infrastructure by which we prosecute and jail criminals can be biased.

Chances are, you live in a place that has a courthouse centrally located in the nucleus of the city near other important civic institutions like the post office, city hall, police station, theater, and other government buildings. Other spaces, like prisons and local jails, find themselves obscurely hidden away on the outskirts of cities and towns and not prominently displayed alongside buildings that serve justice in the heart of the downtown area. A 2017 report from the Vera Institute of Justice cites a substantial increase in the construction of new jails in rural areas, far from the cities and towns in which infractions are actually committed.[35] And according to the Prison Policy Initiative, "Since 1980, the majority of new prisons built to accommodate the expanding U.S. prison population have been placed in non-metropolitan areas, with the result that the majority of prisoners are now housed in rural America."[36] Generally speaking, the typical spatial blueprint for an American town is to have justice on full display at the center of the population and the prisoners shamefully cast outside of it. The physical location of the courthouse can even encourage or discourage the use of the building as a public gathering space for assembly, protest, or media coverage.

Courthouses – where justice and freedom are won and lost – often are the architectural crown jewels of the city, not unlike this U.S. courthouse that takes up an entire city block in the historic downtown riverfront of Wilmington, NC (the nearest courthouse to me).

Figure 5.2. The foreboding U.S. Courthouse in Wilmington, NC.

Its presence is imposing on anyone who lives, works, and visits this small southern town. The building sits perched along the Cape Fear River, a strategic land grab by town planners in 1819, who required duties from ships entering and leaving the Port City.

Although Wilmington, NC, was incorporated in 1739, this particular courthouse was built much later, in 1919, a third iteration of previous courthouses and other government buildings that had been lost to fires or had been demolished for renovation purposes. At the time of its completion, the local press described the courthouse as having "a beauty and charm seldom found in connection with a public building."[37] Its symmetrical design and neoclassical architectural style are common in American courthouse design; as explained by the North Carolina Administrative Office of the Courts, "After America became independent from Great Britain, our leaders sought new styles of art and architecture to express the concept of democracy. They began to more closely consider the ideals of government and society exemplified in ancient Greece and Rome."[38] The aesthetics of the building are deliberately meant to communicate fairness and equitable treatment, but what actually happens can be a different story, starting with the entrance into the building.

The "perp walk" is a controversial practice of law enforcement where an arrested suspect – usually handcuffed behind their back – is led from a police vehicle into the jail or courthouse. Media often take this opportunity to photograph the accused perpetrator, or "perp." Previous court cases have established the perp walk as a violation of the suspect's right to privacy, but this practice continues, particularly for cases with high-profile media coverage. The purpose of the perp walk is twofold: (1) to publicly shame the arrested, and (2) to generate positive publicity for law enforcement.[39] Nat Hentoff, a columnist for *The New York Times*, wrote, "Under such circumstances, even Mother Teresa would look extremely suspicious, especially if her hands were cuffed behind her back."[40] The most infamous perp walk in history was probably that of Lee Harvey Oswald, the man accused of assassinating President John F. Kennedy in 1963. Just hours after the assassination, Oswald had already been arrested, processed, interrogated, and detained. Dallas Police Officers were perp-walking Oswald down a crowded hallway when he was shot and killed. Similarly, in 1984, an accused child molester named Jeff Doucet was being perp-walked by police when he was shot in the head by the victim's father, who had been lying in wait. Aside from being a dangerous opportunity for vigilante justice, the perp walk is an unethical and unfair treatment of a suspect with presumed innocence. Designing a courthouse with a private, protected entrance for high-profile defendants would be a simple way to uphold the democratic ideals communicated by the building's architectural façade.

A more ethical design of courthouses should take into consideration more than just criminals, victims, and their families and loved ones. The architecture also should benefit law staff and employees, who spend their professional lives inside buildings where fateful decisions are made (sometimes, quite literally, life and death). With this in mind, the global design firm, HOK, is one company that is committed to designing more ethical courthouses of the future, which recommends the following five themes:[41]

1. Design for flexibility.

   Example: Utilize modular spaces that can be modified for various uses of public spaces.

2. Design for technology.

   Example: Make courts more accessible for virtual participation – court appearances, online jury screening, etc. – while ensuring seamless integration of technology so that it does not disrupt court proceedings.

3. Design for safety and security.

   Example: Create intentional pathways for judicial staff, defendants, and the public to mitigate safety threats. Include security measures throughout the spaces in unobtrusive ways while maintaining transparency that the space is being surveilled.

4. Design for health and wellbeing.

   Example: Incorporate trauma-informed designs that are mindful of anxiety, fear, and other emotions that accompany the legal process. Welcome features like natural light, outdoor views, and open staircases have been shown to have a healing effect.

5. Design for social and environmental equity.

   Example: Use green-building and sustainable building practices and allow for alternative uses of the building to allow for more positive associations with the courthouse space.

Courthouses often reside in the center of cities and towns not simply for their accessibility but because they represent a source of civic pride. They serve as a reminder that we are a civilized people who hold each other to a code of conduct. For these reasons, we must constantly reevaluate the designs of our courthouses (their location, aesthetics, and layout) to ensure their ethical positivity, given that these institutions represent some of the basic most human-centered values of fairness, equity, freedom, guilt or innocence, reckoning, and exoneration. Just as we should not use predictive policing tactics to predict crimes, we should not let our halls of justice predict that justice.

Ethical prison design

John Howard (1726 – 1790) was an eighteenth-century English philanthropist who would later become known as "the prisoner's friend" because of his devotion to early prison reform. In 1773, after being appointed as Sheriff in Bedfordshire, England, Howard began touring hundreds of prisons across Europe. He was appalled at the inhumane conditions of prisons and the unethical treatment of the inmates within, inspiring him to publish his observations in a 1777 book, *The State of the Prisons.* In it, he articulates specific recommendations for improving prison designs and practices for the betterment of the inmates, calling for more sanitary living conditions, a healthier diet, and the general well-being of prisoners. Howard later would present these findings to Parliament, which resulted in legislative changes to the penal code but perhaps more importantly, increased public consciousness of the treatment of the incarcerated population. The result was a humanitarian-led effort that continues its work today; the Howard League for Penal Reform remains the largest criminal justice reform organization in Europe, sparking multiple offshoots like The Howard Association in the United States and the John Howard Society in Canada.

Howard did express concern that with reformed prisons, "the lower classes of people will find them more comfortable places of residence than their own houses" and vowed that incarceration should not be "pleasant."[42] The man who would become synonymous with prison reform also advocated for hard labor and solitary confinement (even for child inmates), adding, "Solitude and silence are favorable to reflection." Although we might find some of his views cringeworthy nearly 250 years later, Howard felt strongly that imprisonment was for rehabilitation – not punishment – and this was reflected in his proposed prison designs, which remain to this day and include open-air exercise facilities, rooms for religious instruction and prayer, medical services, tiered systems of prisoners categorized by the severity of their offenses, and the single prison cell, which afforded personal space and privacy. John Howard's efforts were progressive and ethically justified at a time when prisoners were mistreated and forgotten. He pleaded, "Let them [the prisoners] be managed with calmness, yet with steadiness; show them that you have humanity, and that you aim to make them useful members of society."[43]

Scholars point out that it was John Howard who inspired the well-known prison design of the panopticon, a western symbol of social control and authority. The panopticon was a circular building design with an observation tower at its center; the outer ring was several stories high and housed rows of single prison cells. This efficient design enabled a single guard at the center of the building to monitor hundreds of prisoners. But, the new prison design was not driven by the cost of guard labor or motivated by the well-being and safety of its inmates. No, the panopticon used a "psychological or social constitutional framework"[44] to exhibit control over its inhabitants via constant surveillance. The central tower obscures the guards within, allowing them to watch prisoners but preventing prisoners from knowing when they are being watched. The architect of this revolutionary and ideal system of imprisonment was Jeremy Bentham, the famous utilitarian philosopher. Whereas Howard focused his attention on the humanitarian design of prisons, Bentham strived for social reform through prison design. Bentham wrote, "Men become at length what they are forced to seem to be."[45] In other words, we behave better when we perceive that we are being watched by others, especially by those who are in a position of authority.

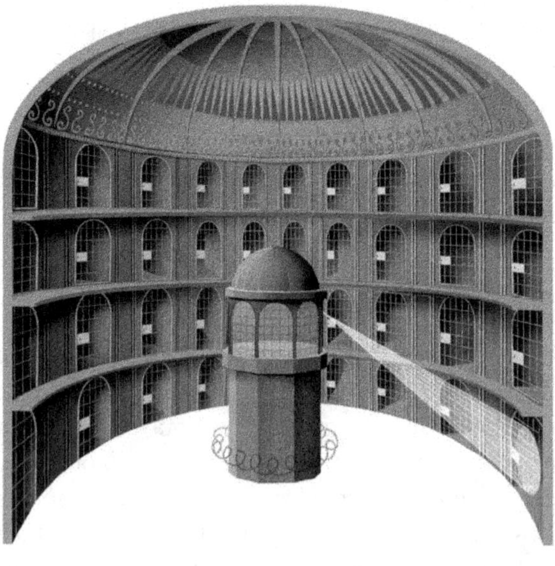

Figure 5.3. Jeremy Bentham's Panopticon embodies architectural control.

Although Bentham conceived of the panopticon design in 1787, the first prototype was not constructed until 1817 in New Delhi. Few panopticon-style prisons exist today, and even fewer remain in operation. A major problem with this omnipotent design is that it assumes that all prisoners deserve unrelenting surveillance (or at least the perception of it) to keep them straight, reinforcing and constantly reminding inmates of the power hierarchy. Panopticon inmates would have found that they had been doubly incarcerated: once for the original crime for which they have been convicted that sent them to prison, and again once they are in prison and presumed to be a dangerous threat unless watched.

Michael Foucault (1926 – 1984) was a French philosopher who termed the phrase "surveillance society." He wrote extensively about Bentham's panopticon and criticized it heavily, saying that its design was a "cruel, ingenious cage" for prisoners. Foucault's objection was less focused on the humanitarianism aspect and more so on the exploitation of power structures. The major impact of the panopticon, he argued, was "to induce in the inmate a state of conscious and permanent visibility that assures the automatic functioning of power... [the prisoner] is seen, but he does not see; he is the object of information, never a subject in communication."[46] The panopticon violates ethical design in several ways. Firstly, there are physically damaging side effects to feeling watched all the time, including increased anxiety levels from fear and uncertainty and physiological and emotional stressors resulting from hypervigilance. Secondly, the watchful design ignores transparency between the guards and the inmates, treating them more as objects than subjects. Foucault raised interesting points about disciplinary control and modern prisons, noting that the primary role of a prison is to deprive inmates of their freedom but not of their personhood.

The ethics of prison architecture are tricky. At a minimum, a prison must be designed for security (for prisoners, for guards, and for the public), but it also must do so ethically and humanely. A meta-analysis of research on prison design has identified two distinct categories of elements that determine the "ethical architecture" of prisons: personal living space and general spaces. Personal living space includes environmental features like lighting, construction materials, aesthetic, noise, views, temperature, air quality, and privacy. The general prison space includes size and crowding, visitation rights, nature, layout, security technology, accessibility, facilities, and the overall age of the structure.[47] Each of these domains should be considered collectively in order to achieve a symbiosis between prison design and wellbeing. However, not everyone thinks that the prison experience should be a positive or even comfortable one. A 1995 article published in the *Tampa Bay Times* titled "Prison Shouldn't Equate to Comfort" opined that "prisons as a vehicle for rehabilitation [is] a largely failed concept" and even lamented that "the concept of disincentive – the role incarceration fills anyway – has become a thing of the past, gone with chain gangs and bread-and-water diets." Others perpetuate the belief that inmates prefer prison to be on the outside; while campaigning for U.S. President, Ben Carson remarked that upon touring a federal prison, he was "flabbergasted by the accommodations" and hypothesized that "a lot of times when it's about time for one of the guys to be discharged, especially when its winter, they'll do something so they can stay in there."[48] This argument is used as a justification to harden punishment and deprive prisons of any semblance of ethical design. But the evidence suggests otherwise; a 2007 study at Harvard University concluded that harsher prison conditions actually boosted rates of recidivism.[49] These findings were echoed in a 2022 study at MIT that analyzed prison conditions and rates of recidivism and found that better prisons that are newer, less crowded, and offer more services and facilities actually reduce the likelihood of re-offending.[50] The public perception is that criminals have it way too good in prison, but the data show that it is actually in society's best interest to treat prisoners well.

Halden Prison in Norway is often given the mantle of "the most humane prison in the world." This is because the Norwegian design philosophy towards jails and prisons is that they are used to rehabilitate, not merely incarcerate. The country has adopted radically humane penal codes that abolished the death penalty in 1979 and life sentences in 1981. In fact, the maximum prison sentence in Norway is 21 years, even for the most hardened criminals like Anders Behring Breivik, who murdered 77 people in 2011 by detonating and explosive and then going on a shooting rampage. According to an expose in *The New York Times Magazine*, "Every aspect of [Halden Prison] was designed to ease psychological pressures, mitigate conflict and minimize interpersonal friction."[51] Starting with the physical surroundings, many westerners would not recognize Halden as a prison because of its natural landscapes, panoramic views, recreational areas, and the perimeter wall, which has no barbed wire or towers with armed guards. Inside the building, there are workshops, classrooms, rooms for religious worship, and even a restaurant-quality kitchen, fully stocked with loads of kitchen tools that could double as weapons. And last, the prison uses "dynamic" as opposed to "static" security measures, in which dangerous or destructive behaviors are mediated by interpersonal relationships with prison staff rather than by non-human techs like remote-controlled doors, surveillance cameras, and fortified barriers like plexiglass between the two groups. To this extent, prison architects intentionally designed the guard spaces to be small and confined so that guards are encouraged to interact more with prisoners. Halden Prison relies on interpersonal relationships – not constant surveillance – to reform – not punish – criminal offenders. Yet despite Norway's progressively humane prison system and its public advocacy for the worldwide abolition of the death penalty, the country's prisons still utilize a pillar of John Howard's idea of prisoner reformation: isolation.

Solitary confinement is the practice of designating a prisoner to a small, closed cell (usually referred to as a Special Housing Unit, or SHU) for 22 to 23 hours per day with little or no external human contact and can be experienced for months, years, or even decades. Human rights attorneys have long argued that solitary confinement practices constitute cruel and unusual punishment, as has the United Nations Committee Against Torture, which has repeatedly classified the use of solitary confinement as torture. The committee also has said that the punitive measure is "contrary to rehabilitation."[52] Numerous studies have repeatedly shown the irreversible damage that solitary confinement has on prisoners, including psychological and mental harm, physical ailments, neurological decrease in brain activity, and even an increased likelihood of premature death.[53] Regardless, the use of solitary confinement is prevalent among American federal and state prisons and jails. Because penal systems are not required to divulge this data, it is difficult to say precisely, but a watchdog group, Solitary Watch, was able to cross-reference data collected from self-report surveys and from the Bureau of Justice to show that on any given day in the United States, there are more than 122,000 prisoners locked down in isolation.[54] And not surprisingly, there is a disproportionate number of Black and Hispanic males in isolation compared to their White counterparts.[55] Solitary confinement is used largely as discipline control but also as an assertion of power. In their book, *Hell Is A Very Small Place: Voices From Solitary Confinement,* the authors describe many documented cases of isolation being used as retribution for non-criminal offenses, such as affiliation with radical political parties such as the Black Panther Party, religious groups such as Muslims who are characterized as terrorists, and even suspected gang members who possess a copy of Sun Tzu's book, *The Art of War.*

Many have asked whether an ethical prison design can even exist, or whether the concept is inherently oxymoronic. Some professional organizations, like the group Architects / Planners / Designers for Social Responsibility (ADPSR), have formally boycotted the design of prisons and jails altogether and have lobbied larger organizations to adopt similar guidelines. The American

Institute of Architects (AIA) is the largest network of architects and design professionals, and in 2020, under mounting pressure from human rights watchdog groups, the organization revised its Code of Ethics and Professional Conduct to prohibit its 96,000+ members from "knowingly designing spaces intended for execution and torture, including indefinite or prolonged solitary confinement of prisoners for 22 hours or more per day without meaningful human contact, for more than 15 consecutive days."[56] Other organizations like the Design Against Crime Research Lab (DACRL) have issued design challenges to improve prison living. The Cell Furniture Project solicited "design solutions for small or confined living spaces" and submissions included innovations like anti-suicide furniture made from corrugated cardboard, ergonomic cushions that help alleviate back pain, and a toilet cover that helps mitigate foul smells (a common complaint among inmates).[57]

Fundamentally, though, the notion of an ethical prison is contradictory. In her book, *The Great Indoors*, Emily Anthes notes that "Most modern American prisons are intentionally harsh places. The buildings are literally designed to punish – to confine and control, shame and stigmatize, dominate and dehumanize." In other words, unethical design is baked right into the prison architecture itself. Supposing that criminal institutions are a necessary component of any society, it appears to me, however, that we can choose our systemic approach to crime and punishment. There is the Hobbesian framework, which views law-breaking as inevitable due to the wicked nature of man, and so an authoritarian state-run system of punishment is warranted fully. Then, there is Rousseau's framework, which views punishment through the lenses of the social contract, public welfare, and social justice. Rousseau wrote: "It is one of the most important items of business for the government to prevent extreme inequality of fortunes... not by building hospitals for the poor, but by protecting citizens from becoming poor."[58] Or analogously, instead of building more maximum-security prisons for criminals, design ways to protect people (especially vulnerable populations) from becoming criminals in the first place. A key requirement of ethical design is that it be human-centered and prioritize well-being.

There is a famous quote that is often attributed to Russian writer, Fyodor Dostoevsky: "A society should be judged not by the way it treats its outstanding citizens, but by the way it treats its criminals." This line has been championed by prison reform groups and activists, but like much of what we read on the internet, it turns out not to be completely accurate. Although Dostoevsky does write about human rights, he never uttered this phrase. However, other notable figures throughout history have made this observation in their own words. Winston Churchill's 1951 book, *Closing the Ring*, included the phrase: "The mood and temper of the public with regard to the treatment of crime and criminals is one of the most unfailing tests of the civilization of any country," and Nelson Mandela wrote in his memoir, "A nation should not be judged by how it treats its highest citizens, but its lowest ones — and South Africa treated its imprisoned African citizens like animals."[59] Misattribution aside, the point remains: our prison structures and practices say a lot about us. So, who do we want to be? Are we ethical or vindictive? Do we seek forgiveness or retribution? Wellbeing or comeuppance?

Case study: Reforming guns

It is a well-known fact that the United States has a gun violence epidemic. Compared to first-world peer countries, the United States ranks first in terms of gun-related deaths (including homicide and suicide), mass shootings, and gun-related injuries.[60] In 2020, firearm-related deaths surpassed car accidents as the leading cause of mortality for American children and teens.[61] Not coincidentally, the United States has the highest per-capita rate of gun ownership. So, we must ask the question: In a country with more guns than people, do more guns lead to more crime? It turns out that the answer to this question is not so straightforward. In fact, I wrote an entire book, *A Philosophy of Gun Violence* (2022), that examined the connection between guns and gun-related outcomes. The book relies on many of the same philosophers I cite in this book — Latour, Ihde, Norman, Gibson — whom I use to argue a phenomenological approach to guns as technological artifacts, showing how their prevalence and their design work together to compel their use and warp the perceived reality of the holder and/or carrier. At the heart of the gun debate, there seems to be a fundamental difference in how people view guns. There is the position of the National Rifle Association (NRA), which holds that "Guns don't kill people; people kill people," and its inverse view, as articulated by Bruno Latour: "You are different with a gun in your hand; the gun is different with you holding it."[62] These viewpoints are at odds with one another; the first implies that the object (the gun) wields no power over the individual and, therefore, only makes us safer against dangerous criminals, and the second implies that the mere presence of the gun distorts our perceived threat and makes our environment less safe. Importantly, the main distinction is that the second view — the phenomenological argument — ascribes agency to the physical object (the gun), which then dictates that there must be an ethical component to its design and to its designers.

Peter-Paul Verbeek has written volumes on this matter, so I will not repeat all of it here. But, in summary, he argues that objects are involved in (or *mediate*) our decision-making, and subsequently, our actions. And because objects are designed and made by humans, these humans become moral mediators, responsible for their use. Guns are, by their very nature, unethical; designer Mike Monteiro points out that, "If a thing is designed to kill you, it is, by definition, bad design."[63] And in the book *Tricky Design: The Ethics of Things*, the authors affirm that "the very narrow purposes of guns are always a 'no' to life," even when used as a justification for self-defense or for recreation such as hunting. They conclude: "The risks to life of guns being present in contemporary society and culture as real, non-human actants cannot be tolerated if we are to recognize the imperative of responsibility towards life."[64] It is the phrase "imperative of responsibility towards life," specifically, that conflicts with the idea that the gun is anything but morally and ethically reprehensible. Setting aside for the moment that, yes, the gun can be used for good or for recreation, ultimately, it is an object designed to kill. Furthermore, the gun intimidates and imposes its will on others – its mere presence is responsible for creating a power imbalance that affects everyone in its wake.

Is there a scenario in which the gun is not unethical? In an article in *The Atlantic*, "The Philosophy of the Technology of the Gun," philosopher Evan Selinger imagines the many ways that guns could be used, while ignoring their intended design to shoot and kill: "Guns can be tossed around like Frisbees. They can be used to dig through dirt like shovels, or mounted on top of a fireplace mantel, as aesthetic objects."[65] Of course, these are uncommon and even absurd uses of a firearm, but these examples demonstrate Ihde's concept of multistability and how the form of an object drives its function.

Many artists have used the materiality of the gun to demonstrate its undercurrent of harm. In the 1971 performance art piece "Shoot," conceptual artist Chris Burden filmed himself getting shot in the upper arm at close range with a rifle. His piece is a commentary on the relationship between violence and control, particularly within the context of war, and the short 8-second video resides in the Museum of Modern Art. In 1974, a performance artist named Marina Abramović stood in an art studio in Naples in front of a table with 72 objects laid out and, including benign objects like lipstick, perfume, flowers, feathers and yarn, as well as potentially harmful objects like a hammer, scalpel, box of razor blades, an ax, metal spear, a gun and a bullet. Over the span of a 6-hour period, Abramović stood motionless and silent with only a placard that read:

Instructions.
There are 72 objects on the table that one can use on me as desired.
Performance
I am the object.
During this period, I take full responsibility.

The performance both challenged the behavior – and the morality – of the audience but also proved the corporeality of the objects and their affordances. A witness later wrote that the performance had begun with "tame" interactions but eventually was stopped because someone had loaded the gun, placed it in the artist's hand and pointed it at her head. Both Burden and Abramović were seemingly prepared to die for their art, which underscored the violence of the gun as a material object and highlighted how its design compels its use. Suppose, then, that we take the inherently destructive weapon and redesign it for a more meaningful, ethically positive use.

Other artists have reshaped the material elements of the gun into something new altogether. Washington, D.C. artist Stephanie Mercedes melts down bullet casings and guns to cast them into musical bells. Mercedes's symbolism is a reference to the "death knells," or the longstanding English tradition of ringing a bell to announce the death of a person. The Mexican artist Pedro Reyes also has turned weapons into mechanical music instruments in his art installation "Disarm" (2013). The collection of handguns, shotguns, pistols, and rifles can be programmed to perform musical compositions, effectively converting a weapon of violence into joyful entertainment. Reyes is well-known throughout Mexico because of his art installation, "Palas for Pistolas" (Shovels for Guns). The artist collected more than 1,500 guns, melted them down, and recast them as shovelheads. Then, Reyes led a public campaign to use the shovels to plant trees in the names of those affected by gun violence. In his own words, "This ritual has a pedagogical purpose of showing how an agent of death can become an agent of life."[66]Similarly, a Colorado-based organization called RAWTools is a coalition of blacksmiths who solicit firearms and forge them into gardening tools like spades and mattocks. Their motto is to "Disarm hearts and force peace."[67] Performance artists resist the argument that the gun is a neutral object that makes us safer. Art fabricators show us that the intentionality of an object and its ethics can be modified and repurposed for good.

At its core, the gun is an unethical artifact because it is designed to wreak havoc on its target. A cursory review of the statistics on firearm-related deaths clearly indicates that guns do not make us safer; they distort our perceived threats, they create a power imbalance for the holder, and they stoke aggression. Yet guns and gun ownership are deeply woven into the fabric of our modern world, and in some countries like the United States, guaranteed as an individual liberty. So rather than debate the merits of whether guns should be owned or even exist, a more practical conversation should center on whether guns should be allowed to assert their influence in our public spaces through prospective measures like concealed carry and arming schoolteachers in classrooms. Although the gun as an artifact remains unethical, there are ethical design elements that might help offset this. For instance, smart guns that use biometric sensors, radio-frequency identification (RFID), and facial and fingerprint recognition technology can ensure that guns are not used by an unauthorized individual. If such a technology were applied to every gun in the United States, it would instantly eradicate all unintentional shootings by children, which, according to the gun violence prevention organization, Everytown for Gun Safety, occurs on average about once per day.[68] Gun rights advocates like to make the argument that any regulation on firearms is an infringement on their personal rights, but upon closer examination, it is the unethical design of the gun that violates our personal safety, sense of freedom, and responsibility to the wellbeing of others.

Chapter 6

Animals

All animals are equal.
Peter Singer

Do animals possess ethics and morality? This philosophical question is not necessarily unanswerable, but it is certainly controversial. Several of the most popular world religions, including Christianity and Islam, do not recognize animals as sentient beings and, instead, place the animal kingdom squarely under the dominion of man. Acknowledging that animals have the same moralities and ethics as humans would pose complications for the ways in which we use (and misuse) animals. However, we run into the obvious issue of not being able to truly know how animals perceive the world since we cannot communicate directly through a common spoken language. We can, though, observe and conduct research on the behaviors of animals, which can tell us plenty about their ethical postures and their capacities for feeling and suffering. The epigraph that was selected for this chapter comes from the philosopher Peter Singer, a notoriously fierce advocate for animal rights. His 1975 book *Animal Liberation* was regarded as an extremely progressive view on the equality of humans and animals, one which Singer himself has compared to Mary Wollstonecraft's early feminist work, *A Vindication of The Rights of Woman*, published in 1792. The Civil Rights movement of the twentieth century addressed similar injustices and demanded racial equality. Singer describes how women and people of color historically have been viewed as inferior to White European men, who have long expected their subservience. Although strident efforts have been made to recognize gender and racial equality, much more social progress is needed. But as equality pertains to animals, many of our modern dietary practices and entertainment and recreational habits are predicated on their exploitation and mistreatment, which contradicts the fabric of ethical wellbeing. Singer coined the phrase "speciesism" to express how humans favor their own interests over the interests of other species, not unlike the scourges of sexism and racism. And while animal welfare has been viewed more positively in recent years, there are still great injustices that need to be rectified.

Acknowledging that all animals have ethical standards – and, by extension, deserve equality – rests on two important assumptions. First, we might disagree on whether animals "feel" in the same ways as we do in the human condition. This can mean emotionally (happiness and sadness), psychologically (fear and anxiety), and even physically (pain and suffering). If we allow ourselves to believe that animals do not experience these same feelings, then it becomes much easier and ethically justifiable to engage in hunting, fishing, or the generally inhumane treatment of animals and pets. In his emotionally powerful song, "Something in the Way," Nirvana front man Kurt Cobain recites the lyric, "It's okay to eat fish, 'cause they don't have any feelings." The line is interpreted as a metaphor for someone who justifies their actions based on a belief that they know to be untrue, since Cobain was an animal rights advocate whose songs often contained references to animal cruelty, which he compared to his traumatic youth experience. Someone who tells himself that an animal is *just an animal* is not beholden to the same standards of ethical behavior, just as there are many examples of the dehumanization of women and of people of color that have provided justification for their mistreatment throughout history. A second assumption is that animal life has "worth," not unlike humans. This concept is a bit more difficult to accept, but as Singer points out, it does not suggest that *all* life (human and nonhuman) is equal. Perhaps one of Singer's most controversial views is the worth that he assigns to human life.[1] In fact, it is Singer's differentiation of life worth based on the capability of self-awareness and consciousness, the prospect of existence without pain, and the capacity for meaningful relationships, among other factors, that allow him to say that the life of a mosquito is not comparable to the life of a Golden Retriever, nor is the life of that dog worth as much as the life of a sentient human being. In other words, "The basic principle of equality does not require equal or identical *treatment*; it requires equal consideration."[2] If we can accept that (1) animals are capable of feeling and that (2) their lives have worth (albeit some lives have more worth than others), then we must re-evaluate our ethical treatment of every species.

People for the Ethical Treatment of Animals (PETA) is an activist organization devoted to the opposition of speciesism, or what it calls the "human-supremacist worldview."[3] The overarching mission of PETA is the expansion of rights to all animals. PETA founder Ingrid Newkirk cites the views of utilitarian philosopher Jeremy Bentham as the basis for the ethical treatment of animals: "The question is not 'Can they reason?' nor 'Can they talk?' but 'Can they suffer?'"[4] Therefore, imposing any type of suffering on animals is morally wrong, and we have an ethical obligation to preserve the rights and wellbeing of animals, just as we do for humans. But following this logic does present some complications. For example, if the animal provides sustenance for a human, and there are no alternatives, doesn't it become justifiable? For better context, does a human who survives an entire winter on the killing and eating of a large-game elk have the same ethical obligation to that elk as a city-dweller in a modernized society who hunts big game for recreation? Or, using another example, is the ritual of animal sacrifice an exercise of religious freedom or a violation of animal rights? As these scenarios show, there are many mitigating circumstances upon which animal ethics depends. Regardless, it is our duty and our ethical obligation to impose, support, and use designs that acknowledge that animals are deserving of ethical treatment and that do not infringe on animal populations that are incapable of speaking out against abuse.

Crucially, this chapter discusses how ethical design applies to animals in cultures where the exploitation of animals is not required for survival or for religious practice. With this in mind, we might expand our definition of *ethics* – "mindful wellbeing" – and the definition of *ethical design* – "that set of standards and practices that guides how an object is shaped and subsequently affects its users as well as those in its periphery" – to include the wellbeing of nonhumans. In addition to the basic expectation that we treat animals with dignity and respect, we also might consider other human behaviors that have profound impacts on animals such as zoos and park designs, conservation and de-extinction efforts, and factory farming, to name just a few.

Zoo design

The largest collection of animals in the world is at the Berlin Zoo and Aquarium, located in the Lichtenberg District in Berlin, Germany and home to more than 20,000 animals from 1,200 different species. It hosts nearly 4 million visitors annually. By comparison, the largest collection in the United States is the Columbus Zoo and Aquarium, which accommodates approximately half as many animals and human visitors each year. In terms of size, The Wilds is located on more than 10,000 acres in Cumberland, OH and is generally considered the world's largest zoo, though it resembles more of an open-air safari park with sweeping landscapes and terrain than a typical city-based zoo. The Schönbrunn Zoo, or Vienna Zoo, is widely recognized as the world's oldest zoo still in operation, first established in 1752 in Austria. What exactly defines a zoo, short for zoological park, is open to interpretation, whether it be for the purpose of education, amusement, or animal conservation. Whether it be in an urban or rural setting, whether it is accredited by a governing body, whether it includes specific types and numbers of species, or whether it conducts its own research. These many features stipulate how we perceive the different types of zoos that exist all around the world and either impart or ignore their ethical justification.

Humans have long been fascinated with exotic animals. The popularity of the modern zoo, where non-domesticated animals were put on display for public viewing, began shortly after the French Revolution (1789-1799). According to Natascha Meuser, author of *Zoo Buildings*, early zoos served the "hybrid role as a prison, theater, and museum"[5] where visitors could observe wild animals in enclosures that resembled more of a prison cell than a habitat. Unfortunately, this unethical and hostile design persists today, as depicted in the image below of a bear enclosure exhibit at the Bitola Zoo in North Macedonia. The pointed iron bars embody the hostile design that effectively deters the bear from escaping its concrete pit.

Figure 6.1. This Bitola, North Macedonia bear enclosure exhibit uses an unethical design.

Conversely, the image below is from the exhibit at the Minneapolis Zoo titled "Russia's Grizzle Coast," which is designed to mimic the rough landscape native to the Russian grizzly bear while still providing a safe buffer between the animals and visitors and safely containing the bears within their restricted but expansive area. The exhibit simulates a degree of freedom for the bears without compromising the safety of the zoo visitors, but it also creates a better context for the viewer to understand how the bears interact within their natural environment.

Figure 6.2. A bear enclosure exhibit at the Minneapolis Zoo resembles a natural habitat.

Meuser describes the evolution of zoo design in five generations, but it was not until the fourth generation during the 1960s that zoo architecture began to replicate the natural environment and attempted to hide restrictive features like barriers, fencing, and operation centers. Many zoos are now in the fifth generation, where they serve both a functional purpose (safe containment of wildlife) while adding education through the exotic architectural design of buildings and surrounding landscapes that are native to the animals on display. Today, zoo visitors learn about the animals and about their heritage and foreign cultures through architecture and park aesthetics in addition to simply observing the animals. However, Meuser cautions that architectural design can sometimes make zoos feel like amusement parks instead of scientific institutions, which is what their primary purpose should be. She warns against the use of "Mickey Mouse-architecture stuff" in zoo design and instead advocates for the balance of the design elements that make zoos equally safe, entertaining, and educational for the public.

Zoo design plays a pivotal role in the wellbeing of its animal inhabitants. Heini Hediger was a zoologist in the 1960s who was among the first to recognize this. Her research emphasized the quality of zoological design over the sheer size of the park. That is, the total acreage of a zoo does not ensure the wellbeing of its animals. Nor do the reproductivity and longevity of the zoo animals necessarily equate to a "successful" zoo. Instead, animals need more. Recent research has shown that providing captive animals with behavioral choices actually lowers their stress levels.[6] Some innovative zoos, like the Woodland Park in Seattle, WA, allow animals the ability to search for and catch their food within their habitat rather than relying on human feeding cycles; the Sydney Zoo in Australia gives large animals like gorillas the freedom to navigate between park enclosures. These modern designs are based on Hediger's five domains, which she developed to measure animal wellbeing in captivity: to achieve competence, to be able to choose, to take control, to experience variety and to engage complexity.[7] The image above from the Bitola Zoo is void of all these domains, whereas the image from the Minneapolis Zoo embraces each of them. The wellbeing of the captive animal depends heavily on the ethical design of its environment.

The design of a zoo can even swing public perception and approval. Make it too safe with highly visible barbed wire, fencing, and harsh structures, and it feels like an animal prison. Make it too kitschy or garish, and it feels like an amusement park. Make it too sterile, and it feels like a museum. And zoos already face strong criticism over their mere existence. Although PETA has been vocal about its opposition to animals being used for entertainment purposes (e.g., SeaWorld, roadside zoos, circuses, and petting zoos), it takes a softer stance on zoos that operate with the central mission of wildlife conservation. The activist organization has been successful in litigating the animal cruelty that is often found in roadside zoos, forcing many of them to be shut down. Alternatively, PETA does support "animal sanctuaries"; the key distinction between the two lies in their design. A pseudo-sanctuary (or fake sanctuary) will use confined spaces and cages rather than natural elements and space to exhibit their animals in isolation. A legitimate sanctuary will design its exhibits for the physical, psychological, and social stimulation of its animals. An article on PETA's website reminds us that "a true sanctuary places the needs and desires of its inhabitants over those of paying visitors and provides a lifetime of excellent care."[8]

The spatial layout of a zoo – both in the design of its animal enclosures and in the larger park design for guests – can even influence visitors' perceptions of animals. Showcasing a particular species at the entrance or focal point of a zoo can garner more attention and potentially more support for public funding for conservation efforts. For example, the highly active sea lions are the first exhibit that visitors encounter at the Central Park Zoo (US) and the Edinburgh Zoo (UK). The bright pink flamingo lagoon ushers visitors into the San Diego Zoo (US) and Bioparc Valencia (Spain). The rest of the zoo park layouts often organize animal exhibits according to their biomes and natural geography rather than their taxonomies or other groupings. This makes navigating the park easier for the visitor, who ultimately is the main audience. *Zoo messaging* can take the form of signage, pathways, location of facilities, and audio ambience that are used to communicate specific information to visitors, usually with the goal of educating people. A 2015 study found that 95% of zoo-goers read at least some of the exhibit signs that provide detailed information about the animal on display.[9] Further, the physical construction and architectural design of animal enclosures and displays can signal messages to visitors, either intentionally or unintentionally. An article in the *International Review of Landscape Architecture and Urban Design* states:

> Putting a single bird into a small cage communicates: 'Here is a beautiful object that we treat as we want and will eventually replace.' In contrast, displaying multiple species of animals in realistic recreations of their habitats communicates a message of both wildlife ecology and appreciation.[10]

Although the wellbeing of animals is a critical aspect of any responsible zoo, the main audience is the visitor. The difficulty in ethical zoo design lies in achieving a balance between the humane captivity of animals and the visitor experience that ultimately helps fund the zoo through ticket sales and donations. Zoo designers utilize the *immersion principle*, which states that "animals and visitors are integrated into recreated biotope or ecosystem scenarios resembling the wild."[11] Even in densely populated urban settings like New York City, a zoo visitor feels as though they have been transported out of the big city and into the African savannah, the Amazon rainforest, or the arctic Tundra. In fact, the design flow of the walkways, paths, and zoo exhibits has been likened to the movie-going experiencing, where visitors are entertained through stimulation and surprise wildlife encounters. An article in the *World Association of Zoos and Aquariums (WAZA) Magazine*, reminds zoo designers and developers that they are "storytellers, creating an emotional bond between guests, wildlife and wild lands."[12]

There is a famous quote that circulates among animal activists, veterinary schools, and social media accounts that read: "The greatness of a nation and its moral progress can be judged by the way its animals are treated." This quote is often misattributed to Mahatma Gandhi, and while it expresses a noble sentiment, there is no direct evidence that Gandhi ever articulated this in any of his speeches or writings. PETA has used the quote in its celebration of Gandhi's birthday, and even Peter Singer cited it in an editorial piece titled, "Moral Progress and Animal Welfare."[13] Though this particular quote seems to be fake, Gandhi did have plenty to say about animal welfare. He wrote this in his autobiography: "I hold that the more helpless a creature, the more entitled it is to protection by man from the cruelty of man."[14] Gandhi clearly understood that the greatest threat to animals is man, and for this reason, it is the duty of man to protect animals, who he said, "are endowed with a soul as that I am."[15] Yet, this is the ethical paradox of zoos. Advocates of zoological parks will emphasize the common good that results from the wildlife conservation efforts, research, education, and public awareness campaigns that zoos produce. Skeptics will question the voyeuristic nature of an institution such as zoos, which sometimes exploit vulnerable and endangered populations incapable of providing their consent to be on display.

The worldwide revenue of all zoos and aquariums topped $22 billion in 2023, with the United States valued at just over $3 billion. This includes zoos, safari parks, and conservatories of all types and sizes. Flatly stated, zoos are big business, and animals are their main product. Although animal activist organizations like PETA recognize that a well-designed and well-run zoo can help with animal rehabilitation and the protection of endangered species, many organizations disapprove of the issue of captive breeding (or the practice of forced mating or *husbandry* to proliferate a species or animal rather than recapturing more from the wild). According to a press release on its website, PETA calls for a shift away "from breeding more animals to keep in cages and tanks, to directing attention to the root causes of why animals all over the world are in peril: habitat destruction, poaching, and the exotic animal trade."[16] The American Society for the Prevention of Cruelty to Animals (ASPCA) takes a more lenient position on captive breeding, only requesting that zoos participate in "tightly controlled breeding programs and take responsibility for all their animals and their offspring, even when they are no longer under their direct care."[17] We must keep a watchful eye on zoological parks to ensure that animals receive the equal consideration that they deserve.

Zoos may look very different in the future. The diffusion of new technologies like augmented and virtual reality holds strong potential for generating a more immersive experience with animals in captivity that requires less interaction between visitors and animals while providing more public education – something that most zoos prioritize in their own mission statements. In his book, *Zoo Studies,* Paul A. Rees theorizes how zoos need to adopt a more politically correct approach to avoid racial stereotyping and cultural appropriation in their exhibit designs, and that zoos can be a proxy for sensitive international relations, as was the case recently with the custody battle over the exchange of pandas between the United States and China.[18] Rees even speculates that zoos could be filled with cloned animals in the future, which will raise a whole new set of ethical questions as to how or whether we should engineer life for our own entertainment.

Necrofauna, or de-extinction

Humans are animals. Among the many features that distinguish humans from other types of animals is that we can manipulate our earthly co-habitants. We place exotic animals in captivity for our entertainment, we domesticate certain animals for companionship, and we genetically engineer animals to suit our palates. And we use non-animal species of flora – plants and flowers – in purposeful ways to serve our desires and our needs. We install lush Bermuda sod on golf courses in the middle of deserts, we plant tropical palm trees in northern latitudes, and we grow almonds, which require vast amounts of water, in the most drought-prone regions of the United States. Often, these exploits are driven by economics rather than ethics, and rarely do we consider the consequences of the latter. Just because we *can* doesn't mean that we *should.*

Biologists typically categorize species of plants and animals as *native, non-native,* or *invasive.* A native species is historically local to that area and is acclimated to the needs of its biome, whereas a non-native species has been introduced to a new region, either by humans or by natural processes. There are simply more benefits to native than non-native species. In other words, Mother Nature knows what she is doing. A native oak tree, for example, is far more successful in providing for native animal species like moths and butterflies, which produce larvae, which provide sustenance for native birds, which pollinate plants, regulate disease, and maintain a healthy ecosystem.

You know, real circle-of-life type stuff. Conversely, a non-native introduced species would not support the environment the same way as a native species and could become detrimental to the ecological cycle. This is the third category – invasive species – reserved for those considered destructive to their surroundings.

Invasive species can be introduced by humans on purpose or accidentally, or they can be naturally occurring. An example of a purposeful introduction of an invasive species is the Common Starling bird. Originally native to Europe and Asia, this songbird was introduced to New York City in the late 1800s by a group of literary ecologists who wanted every bird that is mentioned in Shakespeare's plays to be represented in the United States. Thanks to The American Acclimatization Society, there are now more than 200 million Common Starlings enumerated in North America, and they are generally considered to be airborne pests that transmit avian diseases, decimate crop fields, and tyrannize other bird species. Sometimes, an invasive species is the result of accidental or negligent human behavior. In Florida, non-native Burmese pythons pose a significant threat to native wildlife. It is widely believed that the tens of thousands of pythons that currently populate the Everglades National Park are descendants of a former pet that was released into the park in the 1970s. Invasive species can also be naturally occurring. Recent research in the field of *invasion biology*, or the study of invasive species in the environment, shows that animals are migrating into unfamiliar geographic regions to escape a changing climate. These "climate refugees" then become invasive by natural means rather than at the hands of meddlesome humans. For instance, in my home state of North Carolina, there has been an increase in the population of armadillos, an armored reptile that is native to South America and has expanded through Central America and the American Southwest as global temperatures continue to rise.

It is very common for human-introduced animal species to become *invasive* and wreak havoc on the surrounding ecological system and its biodiversity. An article in the scientific journal, *PNAS*, reports that 1 out of 4 vertebrates that are deliberately introduced into an environment eventually become invasive and that once they have become established, the introduced species is nearly impossible to control.[19] There are countless examples of efforts to eradicate invasive species, almost always to no avail. Regarding the Common Starling bird that was introduced by the Shakespearean group, the USDA Wildlife Service poisoned 790,000 of them in 2020 alone.[20] Similarly, efforts to eliminate the Burmese python from the Florida Everglades is an ongoing effort; the state

Fish and Wildlife Conservation Commission encourages the public to "humanely kill" pythons in the wild and even sponsors the ten-day Florida Python Challenge®, which offers a $10,000 grand prize and $7,500 runner-up prize for the most pythons captured and killed.[21]

In addition to the environmental cost of invasive species, there are economic liabilities. The cost of invasive plants and animals has been estimated at $100 billion in the United States alone.[22] A 2016 report from the U.S. Department of the Interior highlights the ways in which invasive species threaten tens of billions of dollars' worth of damage to the built environment, in particular, the four main areas of infrastructure: power systems, water systems, housing systems, and transportation systems.[23] Some places, like the island nation of New Zealand, have taken extreme measures to try and address their invasive species problems. The *Predator Free 2050* project was conceived in 2016 by the New Zealand government and endorsed by local conservation groups to completely eradicate three of the island's most damaging non-native mammals: rats, stoats, and possums. The ambitious project is estimated to cost the country $1 billion per year.[24]

Very rarely does the attempt to control an invasive species or the introduction of a non-native species result in positive outcomes. Through our efforts to do so, we are constantly reminded that nature is a powerful force over which we have little control. Often, the solution to regulate any unwanted species or attempt at population control is through total eradication. But as author Mary Roach writes in her book, *Fuzz: When Nature Breaks the Law*, "Natality is more effective than mortality." Rather than cull through populations of invasive species, we can be more cognizant of our actions and instead adopt preventive measures and designs in our behaviors and built environments to co-habituate with native species. Roach interviews a spokesperson from the National Wildlife Research Center, who says, "When it comes to wildlife issues, [it] seems like we've created a lot of our own problems." Perhaps we should heed this caution as we embark on the new frontier of genetic engineering.

In 1982, researchers George Poinar and Roberta Hess published the article "Ultrastructure of 40-Million-Year-Old Insect Tissue" in the journal *Science*. The paleobiologists argued that amber, the sticky resin that oozes from tree bark, could preserve molecular structures like nuclei and mitochondria in once-living organisms (such as a mosquito). And, if ancient amber could be recovered from a prehistoric forest, then we might be able to examine a well-preserved organism that had fallen into the gooey amber millions of years in the past. If this hypothetical scenario sounds familiar, it is because this paper inspired Michael Crichton's bestselling book *Jurassic Park*, which would later become adapted into a major motion picture of the same name. What if we could resurrect animals that once roamed the planet? What if we could hack the process of natural selection and override its evolutionary processes? Setting aside the scientific plausibility of preserving DNA in amber, the paper and subsequent book then film marked the beginning of a fascination with *necrofauna*, or the process of de-extinction. The International Union for the Conservation of Nature (IUCN) defines de-extinction as "the ecological replacement of an extinct species by means of purposefully adapting a living organism to serve the ecological function of the extinct species."[25] The key word *purposefully* refers to the conscious, deliberate effort by humans to reinstate a living organism back into the ecosystem for a specific purpose; therefore, necrofauna is a design choice for which we should evaluate its ethics.

Through necrofauna, scientists can leap over natural selection processes and intentionally design a species back into the natural environment that had once eliminated it for one reason or another. Indeed, there have been numerous attempts at de-extinction. In 2003, Spanish researchers used a cloning technique to reanimate the bucardo, a species of mountain goat that had gone extinct just three years earlier. The experiment was successful but short-lived, however, as the bucardo survived for only seven minutes. The project received widespread praise and criticism; scientists in extinction studies heralded the technological breakthrough, whereas others raised serious concerns about the ethical, physical, and financial costs of de-extinction. In a 2018 paper published in the *Journal of Agricultural and Environmental Ethics*, philosopher Heather Browning identifies a series of issues related to the revival of extinct species, including whether the modern ecosystem can sustain them, whether they would pose a significant threat to humans and/or other animals, the re-allocation of funds towards de-extinction projects and away from current conservation efforts for endangered species, as well as the moral and ethical conundrums that accompany these concerns. After careful consideration, Browning concluded that "none of the proposed benefits to de-extinction programs appear sufficient to outweigh the cost in terms of animal welfare, at least not as it currently stands."[26]

Nevertheless, efforts persist to achieve the goal of de-extinction. Specifically, the Dallas-based company, Colossal Biosciences, was founded in 2021 by Harvard geneticist George Church and has pledged to revive three species: the Wooly Mammoth, the Tasmanian tiger, and the dodo bird. So far, Colossal has raised more than $150 million to bring the Wooly Mammoth back to life by the year 2027.[27] Colossal justifies its mission to re-animate the ancient animal in a promotional video, saying, "Reintroducing wooly mammoths to their former homes in the Arctic regions will help to bring balance back to their habitat and slow or even reverse the damage that climate change has done to our planet"[28] But scientists point out that this environmental argument is a thin one at best. Due to climate change, the natural habitat of Wooly Mammoths is significantly warmer now than at the time of their existence and would require an immediate adaptation from the animal that last roamed the earth during the ice age. And the re-introduction of any extinct species would only contribute further to the methane emissions that accelerate the warming of the planet, which we are furiously trying to counter. Although there are some environmental benefits to reviving the Wooly Mammoth, these are easily outweighed by their negative impacts.

Obviously, necrofauna raises many philosophical, legal, moral, and ethical concerns. And to be honest, I'm a bit ambivalent about it. Some scholars have argued that the reanimation of species like the Wooly Mammoth, the dodo bird, the carrier pigeon, and hundreds of other animals is morally permissible because their extinction was caused originally by humans.[29] Under this moral imperative, de-extinction is not irresponsible; it's justified. To me, this deontological, restorative justice argument of "owing it" to a species that we caused to go extinct is rather shallow, as it overlooks the ripple effects that it would cause to future humans, other animals, and the environment which we would then "owe" even more reparations. The moral imperative argument does seem like circular logic. That said, I could understand the de-extinction effort if it had a more ethical purpose, something that the example of the Wooly Mammoth seems to lack. George Church – the founder of Colossal Biosciences – has said that the revival of the Wooly Mammoth could be a boon to the tourist economy and other "business models" such as the procurement of mammoth meat, hair, and ivory.[30] The profit motive should not be justification for de-extinction. However, there are some instances where de-extinction could be ethically permissible, such as in cases where an endangered species is on the verge of extinction, where animals hold significant religious or cultural value, or where certain species can rehabilitate an ecosystem and contribute positively to biodiversity. Many people are unsure what to think of de-extinction; it is exciting to think of the possibilities but frightening to think of the repercussions (for this juxtaposition, consider how the viewer feels at the beginning and at the end of the film, *Jurassic Park*). Britt Wray is the author of *Rise of the Necrofauna: The Science, Ethics, and Risks of De-Extinction*, and she feels similarly conflicted about what she terms "human-mediated nature." She reminds the reader that de-extinction is a fascination, not a necessity, and that "de-extinction is still just an option, as there will always be a human at the helm who decides whether it will go forward or not." Her conclusion cannot be understated. Ultimately, necrofauna is a *design choice*, as it pertains to the purposeful arrangement of our environment, and we must seriously evaluate the ethics behind that choice.

Factory farming

So far in this chapter, we have seen how ethical design choices affect animal welfare. For example, it is not that zoological parks or necrofauna (de-extinction) are inherently unethical concepts. In fact, it can be argued that both can serve noble purposes in some cases. But what determines this is the execution of *design* in their use. A zoo designed with animal wellbeing in mind is ethically permissible, just as de-extinction can be an ethically positive cause. Conversely, zoos and de-extinction projects that exploit animals rather than help them become quickly unjustifiable. The focus of this sub-chapter – factory farming – is an example of how the design of a particular system can determine its goodness. To be clear, it is not my argument that eating meat is unethical – it is the system design of the meat-producing industry that crosses ethical boundaries. But first, let's explore what factory farming entails.

Factory farming is an industrialized process meant to maximize the efficiency of producing meat, eggs, and dairy from caged animals. The United States Agriculture Department (USDA) refers to factory farms as Concentrated Animal Feeding Operations, or CAFOs. To qualify as a CAFO, the farm must have at least one of the following: 125,000 chickens, 82,000 hens, 2,500 pigs, 1,000 cows, or 700 cows. In other words, this is not your typical family farm. According to *Animal Equality*, each year, there are more than 72 billion land animals killed – and the United States accounts for about 10 billion of these animals (9 billion of which are chickens). CAFOs provide 72% of the world's poultry and 55% of the world's pork. Globally, 90% of farmed animals are factory farmed. CAFOs are especially predominant in the United States, where 99% of meat consumed by Americans comes from factory farms. This is due in part because CAFOs are precisely *designed* to be large-scale slaughterhouses that mechanize killing and are financially incentivized to produce maximum quantities of meat, eggs, and dairy to be packaged and placed on the shelves of every major retailer and grocer. While a pure capitalist might see nothing wrong with this model, there are clear ethical tradeoffs for the animals, for the laborers, and for the environment.

In their book, *Why Animals Matter*, authors Erin Williams and Margo DeMello give a vivid description of how the factory farming process works. They describe how animals are not raised but instead brought into existence simply for the purpose of dying. Williams and DeMello explain that animals on factory farms "often have no fresh air, do not feel the earth under their feet, and do not enjoy sunlight...They can hardly perform any of their most important natural behaviors or experience even the most basic pleasures."[31] Factory farm animals are alive but deprived of living. This is particularly troubling, since it is widely agreed upon (though it once was not) that animals can feel pain and suffering, just as humans do. Williams and DeMello go on to say that:

> Animals in factory farms are stripped of their "animalness" – their ability to reproduce on their own, to raise their young, and, in many cases, to socialize with their own kind. They are denied pleasures of running, sleeping in the sunlight, and playing. As the animals are turned into profit-producing objects, we are stripped of some of what makes us human: our sense of right and wrong, and our ability to show compassion to those who are less powerful than we are.[32]

There are many unspeakable atrocities that occur on factory farms – many of which have been documented by filmmakers, authors, investigative journalists, and animal rights watch groups. Some of the more inhumane contraptions include gestation crates for pigs and battery cages for chickens. A gestation crate is a metal apparatus meant to restrict the movement of a single pig (or sow) for the duration of a gestational period, or about 4 months. The figure below shows how gestational crates completely immobilize pigs, preventing them from lying down, turning around, and interacting with other pigs. Restricting pregnant pigs to gestation crates ignores animal welfare, as pigs are highly active and socially complex, but it increases the likelihood of a successful pregnancy – and, in turn – profitability.

Figure 6.3. Pigs in unethically designed gestation crates.

Some countries, including the United Kingdom, Switzerland, and Sweden, have banned the use of gestation crates altogether. In the United States, 10 states have passed legislation to prohibit gestation crates, though this only accounts for roughly about 6% of domestic pork output.[33] Because of their cruel design, many retailers have boycotted meat producers that use gestation crates; Chipotle and Whole Foods were among the first companies to make this pledge in 2002 and 2003, respectively, and now most major fast-food chains, grocery stores, and professional food service companies have adopted similar policies.

There are other hostile designs used by CAFOs meant to maximize efficiency and productivity. For instance, battery cages are stacked metal units that house small to medium-sized animals, including rabbits, quails, and, most commonly, egg-laying chickens and hens. They come in many different shapes and sizes (e.g., A-type and H-type designs), but essentially, battery cages are used to segregate poultry and other animals into smaller housing units. Much like the pig gestation crates, the benefit of battery cages is the control over the "incident rate" of poultry to ensure that they live only to generate product (eggs) and nothing more. The sloped design of the cage allows the freshly laid eggs to roll down to the front for easy collection by a farm worker. The number of poultry housed in each unit varies, but in the United States, laws require that each bird has 67-87" of square space: an area smaller area than a single sheet of paper. This cage design does not even allow for the chicken to spread their wings fully.

Figure 6.4. Battery cages maximize space on a factory farm.

Several countries have passed legislation banning the use of battery cages, including Germany, New Zealand, Norway, Switzerland, and others. But in the United States, battery cages remain legal and are used in roughly 85% of egg production.[34] Though hard to imagine fully, the life of a factory farm animal is an abhorrent existence. It is important to remember that structures like gestation crates and battery cates are unethical designs, conceived of by human designers who are motivated more by profit than animal welfare. And on factory farms, animal wellbeing is not the only collateral.

During the 1900-1910 decade, the United States saw its sharpest growth rate in population.[35] Around this same time, savvy businessmen and entrepreneurs recognized that more people equated to a greater demand for food. Just as Henry Ford's assembly line fed the demand for more vehicles, the food production industry revolutionized its processes to slaughter, package, and distribute its products to a hungry public more efficiently. In his book, *The Chain: Farm, Factory, and the Fate of Our Food*, Ted Genoways tells the origin story of George A. Hormel, the founder of the Hormel Foods Corporation, which still operates today and is valued at an estimated $17 billion. Hormel's small food processing plant in Austin, MN, started in 1893 with one butcher, who slaughtered three animals per day. By 1929, the plant was processing more than 4,000 hogs per day when George retired and handed the reins of the company over to his son, Jay. For comparison, the largest hog processing facilities in the United States slaughter roughly 20,000 hogs per day.[36] The Hormel company even invented several products, including "Canadian Bacon" and "Spam," to not waste any part of the animal carcass and to maximize profit. But, this exponential increase in productivity meant a more hostile work environment for laborers, who were expected to work longer hours, performing monotonous and sometimes physically exhausting tasks for little pay. In his 1906 novel, *The Jungle*, Upton Sinclair exposed the unsanitary and horrific working conditions in Chicago meatpacking plants, where he had gone to work undercover. He writes:

> There were high squeals and low squeals, grunts, and wails of agony; there would come a momentary lull, and then a fresh outburst, louder than ever, surging up to a deafening climax. It was too much for some of the visitors--the men would look at each other, laughing nervously, and the women would stand with hands clenched, and the blood rushing to their faces, and the tears starting in their eyes. Meantime, heedless of all these things, the men upon the floor were going about their work. Neither squeals of hogs nor tears of visitors made any difference to them; one by one they hooked up the hogs, and one by one with a swift

> stroke they slit their throats. There was a long line of hogs, with squeals and life-blood ebbing away together; until at last each started again, and vanished with a splash into a huge vat of boiling water. It was all so very businesslike that one watched it fascinated. It was pork-making by machinery, pork-making by applied mathematics. And yet somehow the most matter-of-fact person could not help thinking of the hogs; they were so innocent, they came so very trustingly; and they were so very human in their protests--and so perfectly within their rights! They had done nothing to deserve it; and it was adding insult to injury, as the thing was done here, swinging them up in this cold-blooded, impersonal way, without a pretence at apology, without the homage of a tear. Now and then a visitor wept, to be sure; but this slaughtering-machine ran on, visitors or no visitors. It was like some horrible crime committed in a dungeon, all unseen and unheeded, buried out of sight and of memory.[37]

This process is relentless. It continues at this very moment, as you are reading this book. All over the world, animals are born into existence, only to experience the inside of a crate, a pen, or a cage, to produce their "yield," and then to meet an early death. Much of this is pointless; in 2019, it was estimated that roughly 18 billion animals were raised for food but never eaten, due to expiration and "sell by" dates. In other words, about 1 in 4 factory farm-raised animals is raised to become food waste.[38] And just as Sinclair wrote appallingly about the laborers in slaughterhouses, conditions have not improved all that much over 100 years later. According to the Factory Farm Awareness Coalition, there are approximately 700,000 factory farm workers in the United States, many of whom are non-English speaking immigrants from low-income households. In fact, roughly three-quarters of factory farm workers in the U.S. are foreign, half of which are not legally authorized to work.[39] These workers experience serious health hazards; long-term exposure to airborne particulate matter can cause serious respiratory and autoimmune diseases, repetitive tasks and motions can induce musculoskeletal disorders like Carpal Tunnel Syndrome, and increased speeds on conveyor-belt processing resulted in an average of two amputations per week in 2018.[40] Despite these risks, the average full-time factory farm worker makes about $23,000 per year.[41] Although Sinclair had hoped his book would make a strong case for socialism, it instead led to public outrage over the mistreatment of laborers, or what his contemporary author Jack London would describe as "wage slavery."

Additionally, there are innumerable environmental consequences of factory farming. Perhaps the most devastating impact of large-scale food production is the greenhouse gas emissions that result from the substantial number of animals that emit vast quantities of methane and nitrous oxide and from the deforestation that is required to grow crops to feed the factory farm animals. Aside from this, runoffs from food agriculture remain the largest contaminator of rivers and streams, and CAFOs reduce air quality overall.[42]

Our current model of meat production has become an ethical dilemma. But the question is not "Is it ethical to eat meat?" The question is, "Is it ethical to support a system of harmful practices by which meat is produced?" There are plenty of things that we can do to combat factory farming. An extreme solution would be to embrace vegetarianism, but a more realistic approach would be to make conscious efforts to scale back on meat-eating. Consumers also can choose products that use cage-free and free-roam practices on their farms.[43] And, of course, we can push lawmakers to regulate the farming industry more closely. In 1958, President Eisenhower signed The Humane Methods of Slaughter Act (HMSA), which required slaughterhouses to "prevent needless suffering" in animals, though this phrase, "needless suffering," really can be applied to the entire factory farm model.

Case study: Fur fashion

When you hear the phrase "animal fashion," what do you envision? Do you think of exotic and expensive clothing like alligator skin shoes or a full-length coat made from mink or chinchilla pelts? These certainly qualify, but so does a leather jacket and goose-down parka. You might also include in that category a cashmere sweater, silk tie or dress, suede shoes, and wool socks. In fact, you could unknowingly be wearing a synthetic piece of clothing or accessory that gets its bright color from natural dyes extracted from plants and animals. When we begin to look more closely, it is quite common for clothing textiles to be animal-derived, even if it is not blatantly obvious. For instance, unless they are consciously using vegan-friendly adhesives, many shoemakers often use animal-based "shoe-glue" made from animal collagen. Animal fashion is pervasive and often undetectable by the naked eye. It even seems that many clothiers are unable to confirm whether their manufacturing processes contain animal byproducts. The unfortunate reality is that the modern fashion industry continues to treat animals as raw material for clothing, shoes, and accessories, just as our Neolithic ancestors did 10,000 years ago.

Animal fur clothing started as a necessity for early human survival in colder climates, but by the eleventh century, fur-wearing had come to signal social prestige and nobility in addition to the warmth it provides. However, by the twentieth century, wearing fur became ethically controversial and socially stigmatized. Animal rights groups have staged demonstrations in front of flagship stores of high-profile designers who use animal furs, stormed fashion show runways in protest of their designs, and some rogue activists even have vandalized celebrities wearing fur in public by throwing red paint on their pelts in multiple occasions.[44] Much of this activity has been associated with the PETA organization, which, in 1990, launched an advertising campaign aimed at bringing awareness to consumers about the mistreatment of animals for their use in the fashion industry. The "Naked Campaign" featured celebrities posing in the nude with anti-fur slogans like "I'd Rather Go Naked Than Wear Fur" and "I'd Rather Bare Skin Than Wear Skin." The campaign officially retired in 2020, and after thirty years of provocative TV and print advertisements, PETA declared that "[wearing] fur is dead."[45] However, it is almost guaranteed that even if a fashion company or clothing manufacturer does not use fur, it likely relies on some type of animal byproduct, unless they are specifically designated as a vegan company that strives to use animal-free materials.

In fact, many fashion companies continue to advocate for the use of fur in their clothing designs. Fashion mogul and former Editor-in-Chief for *Vogue,* Anna Wintour, has defended the use of animal fur, arguing that it is more sustainable and less environmentally harmful than faux fur. Others have provided ethical justification for their personal use of animal fur; comedian Joan Rivers stated of her sable coat: "If an animal had to die to keep me warm for 18 years, it's ok."[46]And the International Fur Federation (IFF) proudly markets its fur-based fashion as "the responsible choice," citing that "the natural fur industry protects and conserves the environment. It promotes the idea of longevity and slow fashion." Its Furmark® program is a global certification and traceable system that certifies the origination, procurement, and selling of each fur-based piece of clothing. Each Furmark® product is evaluated independently throughout the supply chain process – from farm to consumer – to promote transparency and to ensure animal welfare and environmental standards are met.

Animal fashion poses an interesting ethical proposition. On the one hand, exploiting animals purely for aesthetic fashion seems completely superficial and unjustifiable. But on the other hand, a responsible, sustainable, and transparent system for using animal products like fur seems like it could be more ethically positive than manufacturing faux furs, which can accelerate fast fashion and have negative environmental costs. In her book, *Situated Empathy*, Susan Thomas defines *fashion ethics* as: "the potential positive impact that the designer, the buying choices, or the production has on workers, consumers, animals, society and the environment."[47] Thomas actually identifies four phases, specifically, during which ethics should be evaluated: the design phase, production phase, pre-purchase phase, and the post-purchase and consumption phase.[48] In other words, the ethical burden must be considered at different stages by three different actors: the designer, the producer, and the consumer. I will not elaborate here on the last two – the producer and consumer – and instead will focus on the ethics of the designer.

An ethical designer of animal fashion is not necessarily one who abstains from the use of animal byproduct altogether. Certainly, it could be argued that the highest threshold for animal fashion is the one that puts animal welfare and wellbeing above all else. However, there is real validity in the argument that some animal-based materials, like furs, are more beneficial than synthetic alternatives. Perhaps the solution is not whether animal-based products *should* be used, but rather *how* they are used. Admittedly, I feel that the International Fur Federation is a bit too enthusiastic about the use of animal fur in fashion, but their Furmark® program might be a step in the right direction as long as it prioritizes transparency and the wellbeing of animals. The ethical treatment of animals ought to be considered during the design phase rather than deferring to the consumer to make ethical choices.

As is the case with the other topics from this chapter – zoological parks, de-extinction, and factory farming – the key to ethical behavior regarding animals depends on moderation. Each of these things can be ethically permissible if the designer's intentions are honorable. A zoo can be beneficial for animal welfare; de-extinction can service conservation; large-scale farming can be operated humanely. Similarly, animal fashion can be acceptable under some criteria. We should not bring animals into existence for the sole purpose of turning them into raw materials. Nor should we permit the mistreatment of animals just as we should not tolerate it for humans.

Chapter 7

Humans

Men have become tools of their tools[1]
Henry David Thoreau

What makes us human? I certainly am not qualified to give an expert answer to this question, but I suspect our ability to make ethical judgments is one of our most special characteristics. Many animals can show love and compassion to one another, but humans are uniquely equipped to use forethought to anticipate outcomes for how others will be treated and retrospection to evaluate the decisions that led to those outcomes. These executive functions are unique to human intelligence. Unlike other animals, we have developed spoken and written languages, we possess a deep consciousness that makes us self-aware, and we have developed complex technologies. But it is this last item – technology – that has presented new challenges to our humanness. Nicholas Carr is an American author who has written eloquently about technology and its complicated effects on the human race. Carr's 2008 article in *The Atlantic Monthly* aptly titled, "Is Google Making Us Stupid?" questioned the technological world that we were creating for ourselves more than 15 years ago, even before the widespread adoption of the smartphone, the ubiquity of the internet, and the emergence of modern social media. Carr took specific aim at Google, which he argued, "suggests a belief that intelligence is the output of a mechanical process, a series of discrete steps that can be isolated, measured, and optimized."[2] In his view, Google was attempting to distill the highly complex nature of human intelligence – the very thing that makes us uniquely human – into a simpler, more formulaic process that can be articulated and understood by all.

As a brief experiment, I googled the phrase, "How do I become a writer?" The search yielded 4.3 billion results; I was presented with images of famous writers, video tutorials, discussion forums, and links to websites that promised actionable steps for how to become a writer. Although Google can provide useful information on concrete writing skills like grammar, punctuation, formatting, brainstorming, and editing, it can't teach me to become a writer. This is a laborious process that emerges from years of reading, writing, seeking feedback, revising, learning, interacting, and drawing from lived experiences that form a comprehensive understanding of how to write. William Zinsser's *On Writing Well* is a classic 1976 text that outlines many of these heuristics for nonfiction writing, but Zinsser himself asks, "Can such principles be taught? Maybe not. But most of them can be learned." In other words, Google can't teach me to be a writer any more than it can teach me to be an NBA all-star. It can describe the things that make a good writer or an all-star point guard, but it cannot transform me into one. Zinsser also anticipates innovations in writing: "We may be given new technologies like the word processor to ease the burdens of composition, but on the whole, we know what we need to know. We're all working with the same words and the same principles."[3] The word processor changed the *way* we write, but it didn't make us smarter writers. Similarly, technological advancements like generative AI will once again change the *way* we write, but it will not make us smarter writers or more enlightened humans.

We easily confuse *information* that has been presented on the internet with true *wisdom*. In fact, we rely so heavily on computing technology to mediate our understanding of the world that social media has replaced sociability, hyperlinks and websites have replaced close reading, and the general architecture of the internet has replaced (to some extent) our ability to store, consolidate, synthesize, and evaluate information. In Carr's 2010 book, *The Shallows*, he warns that "when we start using the Web as a substitute for personal memory, bypassing the inner processes of consolidation, we risk emptying our minds of their riches" before ultimately concluding, "Outsource memory, and culture withers." [4] The more we invest ourselves in technology, the less human we become. And for some, this is the goal.

The *Encyclopedia of Applied Ethics* defines *transhumanism* as the "position that human beings should be permitted to use technology to modify and enhance human cognition and bodily function, expanding abilities and capacities beyond current biological constraints."[5] We have arrived at a point in human existence where humanity and technology have become so intertwined that we soon may not be able to distinguish between the two. In his book, *To Be A Machine*, Mark O'Connell writes that transhumanism is a "liberation movement," intent on separating humans from natural biology. But in liberating ourselves from biology, he argues, we enslave ourselves to technology. O'Connell concludes that the transhumanism movement is contradictory by nature and only succeeds in succumbing to the "formative pressures on the culture of Silicon Valley."[6] In fact, the World Transhumanist

Association is located conveniently in the tech capital of the world, Palo Alto, California.

We often compare humans to technology: our brain is wired like a computer with memory and storage; our central nervous system sends electrical signals throughout our body like circuitry; our body requires sleep to be recharged every day. But, in fact, we are not computers. What distinguishes humans from computers is that we can engage in reasoning; computers can only simulate this. Furthermore, even the most sophisticated computer is human engineered, programmed to execute functions that have been written by other humans who bring with them their own biases, ethics, values, and morals. As Carr points out in *The Shallows*, we have always worried about how technology will make us less human. Socrates feared that the act of writing would produce forgetfulness, Gutenberg's printing press would blur the lines between reality and fiction, and the internet would overload our brains and make us dumb. But one could make the argument that because humans developed these technologies, then this, by default places all technology under the purview of "natural," and all that exists in the natural world is natural. Therefore, a transhuman cyborg would be no less human than a naturally born human. After all, if someone uses a prosthetic leg, we do not consider them less of a human than someone with two functioning legs.

The Ship of Theseus – also referred to as Theseus's Paradox – is an ancient philosophical exercise that asks the metaphysical question: When does an object cease being itself? In this fictional scenario, the Greek philosopher Plutarch considers a large wooden ship in decay. Its owners slowly replaced each of the wooden planks on this ship with stronger timbers. Once all the wood had been replaced, Plutarch asks, is this still the ship of Theseus, or is it a different ship altogether? There have been many variations of this paradox, including a pocketknife whose blade and handle had been replaced, an axe whose wooden handle and hatchet had been restored, and a hole in a sock that had been patched so many times that the sock was now made up of all new fabric (this last example came from the philosopher John Locke).[7] We might even extend the exercise to our own bodies; given that we shed cellular materials like hair, skin, and fingernails, and that we physically and mentally grow as we age, we can ask: Are we the same person that we were at the age of two? Five? Thirty-five? Like most philosophical riddles, there is considerable debate about the views that we (and other objects) are and are not the same as their original identities. As I am not a trained philosopher, I do not have the answer, nor do I know enough to make a strong argument for either case. What

I do know, however, is that in all the above scenarios, there is an element of *design.* For the wooden ship, pocketknife, and axe, there is a material design that is overtly selected because of the superior affordances that the replacements offer; new timbers strengthen a ship, sharper blades make for a more effective knife and axe, and new fabric creates a more resilient and insulated sock. In the case of humans, we age according to a natural design that has been refined through many millennia of biological adaptation and evolutionary processes. As humans play a greater role in the design of life (and in death, as this chapter discusses), we might consider the underlying ethics that redefine what it means to be human.

Genetic design

Attempts to design humanity are not new. *Eugenics* is perhaps the crudest method for engineering the human species, where reproduction is manipulated to design out undesirable characteristics and promote more desirable ones. Of course, the terms "undesirable" and "desirable" are relative to whomever is doing the designing. Different cultures define *beauty* in different ways depending on age, gender, economic status, tradition, and geographic location. And if we use the American Psychological Association's definition of racism, which is the view that some racial groups are inferior to others because of their distinctive characteristics, then the eugenic theory is overtly racist because it replaces physical characteristics that are deemed to be inferior with one that are perceived to be superior. Sir Francis Galton, the cousin of Charles Darwin, is credited with popularizing the idea of eugenics in the late nineteenth century. Galton wrote voraciously about the subject; his books, *Hereditary Talent and Character* and *Hereditary Genius* both attempt to use scientific reasoning to advocate for the use of "planned breeding" for "improving stock"[8] in humans. Yet, Galton's pursuit of an ideal race ignored his own biases, and it was not because of sheer coincidence that his definition of beauty just so happened to look a lot like himself: White, European, and from a privileged social class. Despite the glaring ethical violations of eugenic theory, many esteemed people throughout history endorsed these same views.

Galton laid the groundwork for others to build on his theory in the coming century; American presidents and foreign leaders, social progressives, intellectuals, and luminaries championed eugenics. Winston Churchill served as vice-president of the first international Eugenics Conference, and he justified the takeover of the American Native Americans in America and the Aboriginals in Australia by a "stronger race, a higher-grade race." President

Teddy Roosevelt lamented how the United States had become "a nation which permits unlimited breeding from the worst stocks" and wrote that "society has no business to permit degenerates to reproduce their kind."[9] George Bernard Shaw, Alexander Graham Bell, Jacques Cousteau, and H.G. Wells all supported the eugenics movement. But perhaps most surprising was the public support of Helen Keller, who was well-known for overcoming her deafness and blindness to become an educator and author, and who defended the decision of a physician to refuse to operate on a physically deformed infant, resulting in her death, citing that the infant's life was "not worthwhile."[10]

Many countries have practiced eugenics in some form or another. By 1907, the United States had adopted its first forcible sterilization laws in Indiana, which rendered "promiscuous" and "feeble-minded" women infertile. This practice remained until the 1980s. From the 1930s to the 1980s, other first-world nations, including Japan, Sweden, Canada, Australia, Norway, and Finland, all enacted legislation that permitted the forced (or coerced) sterilization of various vulnerable populations like racial minorities, mentally disabled persons, and those suffering from certain illnesses. And most horrifically, in the 1930s and 1940s, Nazi Germany enacted the largest eugenics movement in world history: the attempted extermination of the Jewish people and of anyone else not of Aryan or Nordic descent. The Nazi regime attempted to control the "racial hygiene" of its population through mass sterilization, racial segregation, and other social policies, but by World War II, the world had begun to question the ethics of genetic engineering.

Eight years after the end of WWII, two biologists – Francis Crick and James Watson – published a groundbreaking article in the journal *Nature*, which described the structure of Deoxyribose Nucleic Acid, or DNA, for which they were awarded a Nobel Prize in Physiology or Medicine.[11] Crick, often referred to as one of the "fathers of DNA," immediately recognized how this scientific discovery might be manipulated by future genetic engineers. He wrote, "It has not escaped our notice that the specific pairing that we have postulated immediately suggests a possible copying mechanism for the genetic material."[12] And so, the researchers felt it necessary to distance their work from the Nazi movement. Crick wrote: "The main difficulty is that people have to start thinking out eugenics in a different way. The Nazis gave it a bad name, and I think it is time something was done to make it respectable again." Similarly, Watson has advised Germany to carry on with genome research but to "put Hitler behind us."[13] Crick and Watson both appear to have held complicated views of race and articulated these sentiments towards the end of their lives. In

2019, James Watson was stripped of his honors and titles held at The Cold Spring Harbor Laboratory, where he had been the Director since 1976, because of his stated beliefs that intelligence is based on racial identity – specifically, that social policies should not be universal because African populations are intellectually inferior to Europeans (for which there is no scientific basis). Similarly, Francis Crick supported the application of what he described as "positive eugenics" but expressed this mostly in the form of private letters with colleagues rather than in speeches or academic publications that would be met with public scrutiny. In a 2003 interview, Crick predicted the following: "I would be astonished if, in the next 100 or 200 years, society did not come round to the view that they would have to try to improve the next generation in some extent or one way or another."[14] And he was most certainly correct, except that it wouldn't take 100 or 200 years; it would take less than a decade.

Each cell in a living organism contains a sequence of DNA strands that determines the development and function of that organism. DNA influences the physical appearance of the organism as well as its behavior (to some extent), susceptibility to diseases, and even general disposition. Stretches of DNA are called *genes*. Genes are responsible for carrying the hereditary information from the parent to the offspring, which is why you might resemble the appearance of a relative but also is why you were born with the instinctual behaviors necessary for survival (e.g., a healthy fear of fire and other natural reflexes that keep you safe). These heritable genes are what make us unique individuals, but they also are what bind us as a species. I mentioned earlier that the practice of eugenics was a "crude" form of genetic design. This is because eugenic theory uses overt practices like sterilization, social policy, and legislation to design conditions under which reproduction can occur and for whom. A more sophisticated and seemingly humane way to engineer a population would be to manipulate people at the molecular level, prior to birth. *Genetic design* does precisely this; in an evolving scientific discipline, genetic designers modify the genetic or cellular structure of an organism (human, animal, or plant) in order to accomplish a planned outcome. In this sense, the field of genetic design is not too distant from any other design discipline wherein a designer applies his or her tools to achieve a desired user behavior.

In 2012, biochemist Jennifer Doudna and microbiologist Emmanuelle Charpentier co-invented the gene editing system, CRISPR-Cas9 or CRISPR for short. This genetic engineering tool allows researchers to modify the DNA makeup of living organisms by searching for, cutting out, and editing snippets of DNA in any given gene. The arrival of CRISPR marked an important moment of reflection for scientists. As with any technology, CRISPR can have positive uses, such as the eradication of genetic diseases or disabilities, or it could be used to genetically design crops to solve famine, food inequality, or even offset the impacts of factory farming on climate change. Simultaneously, CRISPR can be used nefariously as an agent of racial eugenics. Doudna has long asserted an instrumentalist perspective on the technology and remains neutral in her assessment of CRISPR's ethical and moral responsibilities. In a 2015 TED Talk, she stated, "All of us have a huge responsibility, to consider carefully both the unintended consequences as well as the intended impacts of a scientific breakthrough." And she reinforces this neutral position in her 2017 book, *A Crack in Creation*:

> Above all else, we must respect people's freedom to choose their own genetic destiny and strive for healthier, happier lives. If people are given this freedom of choice, they will do with it what they personally think is right – whatever that may be.[15]

Referring to the ethical theories I presented in Chapter 2, Doudna is very clearly aligning with the "Rights Lens", which states that "Ethical action respects the rights of all, including the right to make your own choices." As I pointed out in that same chapter, however, the right of an individual to choose does not always serve the common good. Instead, we might think of this scenario through the lenses of deontology and utilitarianism. Should we focus on the *intent* of CRISPR gene-editing technology (deontology) or the *results* that stem from its use (utilitarianism)? Regardless of your view, it is not ethically responsible to unleash a powerful tool such as CRISPR and then eschew its potential for disastrous outcomes while using "freedom of choice" as its justification.

Gene editing can very easily resurrect and even bolster the eugenic movement, which would undoubtedly affect vulnerable populations in disproportionate ways. In his book, *Hacking Darwin*, Jamie Metzl writes that "If enhancements are allowed and unequal access continues over time, each generation of an advantaged family could become more genetically enhanced than their disadvantaged peers until the difference between two groups becomes unbridgeable."[16] It seems there is little more than semantics and a very fine line that separates gene editing from eugenic theory. In essence, both seek to perfect biology – however that definition of "perfect" might be perceived. If the belief that some races are "better" or superior to others persists, then gene-editing would likely be used as a tool for "race betterment" (a term coined by Sir Francis Galton).

In reality, many scientists argue that race is a social construct only, and that we should focus instead on ethnicity. The former wrongly implies biological differences between groups of people, whereas the latter distinguishes groups of people based on their histories, languages, beliefs, and traditions. Fearmongers, racists, and propagandists could weaponize gene editing technology using their misunderstanding of science as its rationale. Indeed, racism is alive and well, but it might not be as visible as it was in the early twentieth-century American South, where it was manifested in the form of burning crosses, effigies, and Ku Klux Klan outfits. Instead, racism flourishes in the digital landscape, which thrives on polarization and outrage. Sometimes, racist views are coded in other languages, such is the case with the term "diversity," which FOX News host Tucker Carlson has said "is not a value"[17] and others, including U.S. presidential candidate Vivek Ramaswamy and Canadian conservative leader Andrew Scheer, have said "is not our strength." Metzl corrects this view by pointing out that diversity is essential to evolution, and that "diversity has been and remains the sole survival strategy of our species." Gene editing threatens diversity and, by extension, the human species.

Just having knowledge of the human genome, let alone editing it, might create ethical issues. For instance, genetic screening is becoming more accessible and affordable as part of a woman's prenatal care. This screening can inform the mother whether her unborn child will have physical or mental conditions, such as Down's Syndrome or spina bifida, presenting her with the decision of whether to terminate the pregnancy. Arming the mother with this genetic information implies that a child born with these conditions is somehow "less than" or unworthy of life. Here, serious ethical deliberations arise, such as to what extent the mother is responsible for giving birth to a child who will require extra care and financial support. Obviously, this scenario is subjective: a single mother with four children would be impacted significantly more so than a mother with a partner who has socioeconomic advantages, adequate resources, and greater access to healthcare. A 2022 Pew Research study asked Americans about their attitudes towards gene editing technology that might be used to "greatly reduce a baby's risk of developing serious disease or conditions."[18] Interestingly, respondents were divided; 30% reported that gene editing would be a "good idea," and the same amount (30%) said that it would be a "bad idea." Collectively, however, most respondents (71%) saw gene editing as being acceptable for use only in the prevention of disease, and 74% deemed it inappropriate to use gene editing to enhance physical attractiveness. The survey also found that 51% of Americans fear that if the use of gene editing becomes widespread, the government will not go far enough in regulating its usage. James Watson was astutely observant of this as well; although he embraced the neutral view that "genetics per se can never be evil… It is only when we use or misuse it that morality comes in," he stressed that genetic-editing technologies should be kept out of state control and "in the hands of the people."[19] Perhaps this is what Doudna means by "freedom to choose their own genetic destiny."

Genetic design is a tricky issue. On the one hand, it can improve our lives by lessening the risk of genetic diseases and by having positive environmental impacts; on the other hand, however, gene editing likely would lead to genetic homogeneity in humans, effectively weakening our species and contributing to our demise. History has given us countless examples of people jockeying for racial superiority, and it doesn't work. And when we look for ways to design ***for*** something, we are, by default, designing ***out*** others.

Border control

The border between the United States and Mexico spans roughly 2,000 miles across a diverse terrain of deserts, mountains, wetlands, and urban areas. Although it is not the longest land border in the world (that would be the border between the United States and Canada), it is statistically the deadliest border. The International Organization for Migration (IOM) reported nearly 700 deaths in 2022, and this is likely a conservative figure, as many fatalities go unreported and undocumented.[20] And although many immigrants arrive at formal points of entry into the United States legally as refugees or asylum-seekers, many more attempt to cross the border illegally. The acting commissioner of the U.S. Customs and Border Protection (CBP) estimates approximately 10,000 illegal migrant crossings per day, most of whom originate from Mexico and Central America, and the total number of arrests for illegal crossings has reached an unprecedented rate, exceeding 2 million in each of the previous two fiscal years.[21] The United States is currently experiencing a historically high number of immigrants who want to live there, whether it be in pursuit of economic prosperity, a reunion with family members, as a refuge from oppression in another country, or a bevy of other reasons why the difficult decision has been made to attempt to cross the border illegally. These crossings carry the risk of detainment and deportation if caught by border patrol agents, as well as mistreatment, abuse, and extortion from human smugglers (known as coyotes) who often make false promises of safe passage across the border for tens of thousands of dollars. These coyotes prey on vulnerable and desperate populations of people striving for better lives.

Immigration attitudes are shaped by politics. A 2022 Pew Research Center study found that Americans were largely supportive of accepting Ukrainian refugees into the U.S., following the Russian invasion of Ukraine.[22] However, this attitude was tied closely to support for the war; those who viewed Ukrainians as victims of an unauthorized invasion were more welcoming of the refugees that it had created. Similarly, public approval of welcoming refugees from Afghanistan following the U.S. withdrawal of troops in 2021 was deeply partisan; liberals who supported the troop withdrawal were far more likely to accept Afghan refugees than the majority of conservatives, who were less supportive of accepting Afghan refugees and of President Biden's withdrawal of troops from the region. Another example of the recent politicization of immigration policy concerns former President Trump's 2017 "Travel Ban" (Executive Order 13769), which prohibited refugees from certain countries from entering the United States. The ban is sometimes referred to as the "Muslim Ban" because it targets five countries, specifically, with majority-Muslim populations. The ban was originally upheld by the Supreme Court and then reversed in an executive order by President Biden. Indisputably, Trump has a history of using anti-immigrant rhetoric; his 2016 presidential campaign began with hateful statements like this one: "When Mexico sends its people, they're not sending their best… They're bringing crime. They're rapists." During the COVID-19 pandemic, his anti-Asian remarks on Twitter fueled a spike in hate crimes against Asian Americans, and more recently, Trump has said that immigrants are "poisoning the blood of our country."[23] Other foreign leaders have expressed harsh sentiments towards immigrants, who often are seeking refuge or asylum from terrorist regimes. French President Emmanuel Macron has echoed the statement, "We cannot accommodate all the misery in the world," first uttered by Prime Minister Michel Rocard in the 1980s. The Prime Minister of Sweden has called immigration "unsustainable," and Turkey's presidential challenger ran on the promise that he would "send all the refugees back" if elected. But this anti-immigration rhetoric is really nothing more than coded language for the aforementioned politicians to play on the fears of others and to disguise their racism, xenophobia, and discrimination against those who are different. However, most economists will tell you that immigration is economically beneficial for any destination country.[24] So why do we fight so hard to make our borders physically impenetrable?

Let's return to the example of the U.S.-Mexico border. This border has come to signify more than what it actually is: an arbitrary, imaginary line that has been modified many times over the years, depending on who claims ownership of the land that it bisects. The border establishes national boundaries – thereby creating nationalism for those within – and positions everyone else as outsiders. In the 1990s, the United States adopted an important immigration policy known as "Prevention through Deterrence," which was a concerted effort to disrupt known points of illegal entry and to reroute migrants "over more hostile terrain, less suited for crossing and more suited for enforcement."[25] The conceptualization of the physical barrier, or "wall," that would separate the United States and Mexico originated in the 1940s, but since then, every American president has contributed to its construction – even the ones who said they would not.[26] Although the border wall has been a bipartisan project for decades, it remains a heavily politicized issue. It is of particular significance to Trump supporters, who often can be heard chanting "Build That Wall" at rallies and campaign stops. The phrase refers both to the physical barrier that prevents illegal entry into the United States and also a metaphorical desire to segregate "us" from "them." In the modern age, borders have come to embody more of a socio-political boundary than any meaningful sort of territorial one, yet this has not stopped nation-states from attempting to strengthen their borders in an effort to keep others out and to criminalize those who enter without permission.

Building a wall is an antiquated and ineffective approach to immigration. As U.S. Representative Henry Cuellar points out, a border wall is "a fourteenth-century solution to a twenty-first-century problem."[27] The U.S.-Mexico border is especially futile because it is not continuous along the 2,000-mile border, and it is comprised of a hodgepodge of structures, fences, walls, and other obstacles that can easily be overcome with basic materials like rope. Furthermore, studies have shown that erecting a border wall has only modest effects on reducing the number of illegal crossings. A 1978 article in *The Washington Post* stated that the problem with man-made barriers is that "most of them don't work, as long as there are human beings intent on going from one side to the other."[28] More than 40 years later, when President Biden was asked whether he thought border walls were effective, he responded flatly, "No."[29] So why, then, do we continue to build miles and miles of expensive border wall? Perhaps it is because the wall is not designed to prevent people from entering, but it is a symbolic reminder of exclusion designed to excessively and cruelly punish those migrants who attempt to enter.

Immigration control measures along the U.S.-Mexico border often are designed to injure, maim, or kill. During the late 1970s, President Jimmy Carter approved the construction of new fencing in densely populated cities along the southern Texas border with Mexico. A reference to the "Iron Curtain" that separated Western and Eastern Europe during the Cold War era, the "Tortilla Curtain" was maligned by many residents on both sides. The fences were designed not to prevent but to harm; they were outfitted heavily with razor and barbed wire, which the fence contractor said could "sever the toes of anyone trying to scale [them]."[30] After violent protests erupted over the inhumane design, President Carter was compelled to remove the razor wire, but the fences remained in place. Then, in the 1990s, the "Prevention through Deterrence" policy was designed to redirect migrants away from traditional entry points and towards the Sonoran Desert, where they would "find themselves in mortal danger."[31] And they do; a lack of food and water is the leading cause of migrant deaths, followed by drowning.[32] According to researcher Nicole Rivera, "This plan was intended to create death."[33] Researchers have shown that the current design of the border wall is directly responsible for an increase in migrant injuries and deaths. One study found a direct correlation between the building of a 30-foot wall in San Diego County and a surge in hospital visits from migrants who had sustained "significant brain and facial injuries or complex fractures of the extremities or spine, with many requiring intensive care and staged operative reconstructions."[34] Other objects have been used to create physical danger for migrants. The image below shows the floating barriers that were installed by Texas Governor Greg Abbott along the Rio Grande to deter crossings as part of Operation Lone Star.

Figure 7.1. Floating barriers on the Rio Grande are designed to inflict extra harm.

If you look closely at the image, you can see that between each barrier is a metal saw blade with sharp teeth. The barrier alone presents enough of a challenge for someone to overcome it, so the sole function of the saw blade is to inflict additional physical harm on anyone who dares try to pass. Estimates vary, but the cost of Operation Lone Star has cost taxpayers roughly $4 billion, and the U.S. Court of Appeals has ordered Governor Abbott to remove the deadly buoys, which have been connected to several fatalities of women and children migrants. At the time of this writing, the jurisdiction struggle between state and federal officials continues, as Texas border agents wish to "hold the line" at any cost, and federal agencies are expressing a need for a more humane and ethical approach to processing the illegal migrants.

In the book *Tricky Design: The Ethics of Things*, authors Fisher and Gamman question what it means to design for wellbeing rather than for safety. Specifically, in a chapter titled "Making Safety, Making Freedom: Problem-Setting, Collaborative Design and Contested Futures," author Shana Agid asks the question: "What might it mean to design *for* 'freedom' or 'well-being' instead of *against* crime?"[35] The whole spirit of ethical design is represented by this single question. Through the communicative power of design, erecting intentionally harmful barriers between two countries intensifies the "othering" of the people on the opposite side and criminalizes anyone who wishes to cross that border without going through the proper channels, which are broken. The physical barrier also ignores special circumstances, such as asylum seekers and refugees who may be in peril. But as Agid points out, the notion of *crime* is highly complex and both social and political in nature. Because we view *crime* and *safety* through the lens of having an inverse relationship with one another, it is presumed that border enforcement is an act of public safety because crossing the border illegally has been labeled criminal. Further, politicians constantly frame migrants as rapists, gang members, drug smugglers, and ne'er-do-wells instead of pointing out the economic benefits that they bring and the diverse cultures that add to the melting pot of America. Labeling migrants as *criminals* is an exercise in "problem-setting," or the practice of setting the terms of what is wrong and what needs to be changed.[36] The border, for instance, has been set as a problem. However, the border is temporally dependent on who owns the land and at what point in history. Not too long ago, Texans would be considered trespassers on Native American and Mexican soil. Instead, we might think of *crime* as an "action which happens to have been labeled unlawful in this particular place at this particular time, according to the prevailing social, political and economic order."[37] Rather than design barriers according to the prevailing law at the time, designers ought to design for people who find themselves in deeply complex social and political situations.

The way that we design our boundaries can influence our perceptions and attitudes of those on both sides. A heavily militarized border, for instance, implies the need for protection from the other side, whereas a more humane and non-threatening design diffuses tension between the two sides. Some designers and architecture firms have imagined the U.S.-Mexico border in radical ways. For example, the artist Rael San Fratello describes border relations between the United States and Mexico as hanging in a "delicate balance," which is the inspiration for his Teeter Totter Wall. The design was the winner of the 2021 International Award for Public Art and leverages the steel slats of the border wall as a fulcrum, requiring that people on both sides of the border work together to make the teeter-totter function. The simple design disarms the foreboding structure as a spatial weapon and turns it into a child's playground. Others have reimagined the border not as a menacing, continuous wall that keeps people out but as an open and inviting space that welcomes immigrants, celebrates diverse cultures, and honors past generations of the native peoples who came before us. The identity of a nation is defined by more than its geographic boundaries; the border manifests our ethics, morals, and values as a country.

Death by design

Designers shape our world and our perception of the world. Whether it is the architectural and infrastructure designs that shape the built environment, the graphic designs of consumer products and advertising, or the UX and UI designs of our personal technologies, design influences everything around us. Or, as an article published in the *Journal of Art & Design Education* writes, "Designers have become significant creators of meaning in everyday life."[38] It is true that we are fully immersed in design from the moment we are born until the moment that we leave this world. And it is an inescapable aspect of living that we will all encounter death. So, if designers play an integral role in life, it can be said the same for death. In a blog post on Medium.com, Paul Bennett, Chief Creative Officer of the design company IDEO, inquires: How might design walk with us to the end of the world?[39] This sub-chapter is going to ask some uncomfortable questions, partly because humans are self-aware of our existentialism, and this makes us generally adverse to the topic of death (especially our own). As well it should be noted that death is viewed differently by different cultures and religions. For many Western religions, corporal death is a doorway to an eternal afterlife; many Eastern religions teach reincarnation or rebirth from death; and the "nonreligious" sects like atheists, agnostics, and nihilists believe that we return to nothingness after death. The reality is that no one knows for sure what happens to us spiritually after we die, but we do know that design can factor into how the living experience death. This subchapter will be split into two sections: designing *against* death and designing *for* death. Both will examine the relationship between death and design through an ethical lens.

Designing *against* death

To illustrate the ways in which we design against death, let's use a routine example: You need to make a quick trip to the grocery store. Presuming the store is far enough away that you need to drive a car, you will hop into your vehicle, which has undergone a litany of safety tests and is equipped with many safety design features like air bags, seatbelts, power-steering, electronic stability control, and more, depending on the age of the vehicle, and drive along the roadways, which are designed to be wide enough to mitigate head-on collisions with other drivers, are graded with a gentle slope to shed standing water, have clearly marked road lines, signage, posted speed limits, reflective markers, and maybe even rumble strips and guardrails. Once you arrive at the store, which is in a commercially zoned area that has passed rigorous building inspections, you will find a bounty of pre-packaged and fresh foods that have been subject to strict FDA regulations and have been stocked on shelves by paid employees who are in compliance with OSHA safety guidelines. These layers of protection work together to minimize the likelihood of harm or death, and they are the reason why we are not intimidated or fearful of the trip. Arguably, the deadliest aspect of this grocery store run lies in your food choices; a steady diet of "ultra-processed" foods can lead to higher cholesterol, higher blood pressure, obesity, and even early death.[40] The point here is that we work strenuously to keep ourselves safe – and by default – extend our lives. For some, however, the goal is not just to live a healthy life but to *never* die.

The quest for immortality is as old as humanity itself. The *Epic of Gilgamesh* is perhaps the oldest recorded storytelling of this quest, written nearly 4,000 years ago on Sumerian cuneiform. It is a story of a hero's journey to find Upnatishtim, the only human to achieve immortality. Upon finding him, he tells Gilgamesh, "There is no permanence." The story ends with the hero accepting his mortality. And then there are the fictional writings of the "fountain of youth," first articulated in the writings of Herodotus in the fifth century BC and then made famous by the sixteenth-century Spanish explorer Juan Ponce de León. Of course, the world's oldest religions all recount some version of immortality in their sacred texts (whether it be in the afterlife or reincarnation). In other words, humans have always searched for some loophole to escape death, and this effort to stave off death persists today with a giant assist from technology.

Over the last decade, there has been an increased effort in research to slow the aging process. Large tech firms like Google as well as smaller start-up ventures, have committed billions of investment dollars into the anti-aging movement. In 2013, Google acquired the California Life Company, or Calico, and has since committed more than $1 billion to its longevity research. Like illnesses, the company views aging as a disease that can be "cured," but no significant breakthroughs have been made. Other projects include Ambrosia, which offers its patients over the age of 35 years old an $8,000 blood transfusion using plasma (blood without the blood cells) from teenage donors. No meaningful effects have been observed. Nectome is another ambitious research project dedicated to "brain-preservation technology" that wants to download a living person's mind in the hopes that it can be uploaded to a computer in the future. What is the catch for this technological resurrection? Nectome is "100 percent fatal."[41] That is, you must give up your living brain in the hopes that it can be implanted into some type of transhumanist cyborg in the distant future. But dying now for the chance at life later seems a bit oxymoronic. Then there are AI companies like HereAfter that employ a biographer to document your memories and stories while you are alive so that you can be digitally regenerated as an artificially intelligent being for the rest of eternity. It seems that achieving immortality might just be possible, but it depends on your definition of what it means to be *alive*.

Most of us are not seeking immortality but instead just trying to live. Prevention through Design (PtD) is a design process by which we anticipate and then defend against potential harm. In building practices, PtD is evident in physical construction like railings and balusters that keep us from falling into a stairwell or over a ledge, and through policies such as electrical codes that dictate the minimum distance that outlets need to be located from sinks, bathtubs, and showers. These safety regulations are designed to protect us from potentially dangerous outcomes. But these designs assume that everyone wants to live, and that death is an unfortunate and undesirable circumstance. So, how do we account for users who are not looking to *avoid* death but actively *seeking* it instead?

According to the Golden Gate Bridge Highway and Transportation District, there are an average of 30 deaths per year that result from people committing suicide by jumping off the Golden Gate Bridge. The traffic level deck height of the national monument reaches more than 220' above the frigid waters of the San Francisco Bay, and taking the plunge means an almost certain death. In fact, since its opening in 1937, there have been more than 1,700 suicide deaths by jumpers. In response, engineers have designed and installed the Suicide Deterrent System: a system of steel cable netting mounted to the outside of the bridge that extends 20' beyond its railings.

Figure 7.2. The Suicide Deterrent System of the Golden Gate Bridge in San Francisco, CA.

While the main purpose of the $224 million project is to prevent suicide attempts, the design also serves as a constant reminder of the tragic deaths. For this reason, engineers intentionally chose the least intrusive design to preserve the scenic vistas and original architecture of the structure. Early ideas for the suicide deterrent began in the 1950s, and prototypes included barbed wire and vertical fencing, which would have compromised the art deco aesthetic and structural beauty of this modern marvel in engineering. And although many people have criticized the project as being unnecessary and too costly, the

Suicide Deterrent System is an example of an ethical design that minimizes the invitation of suicide while preserving the original design of the bridge. To be clear, the bridge itself does not create suicidal tendencies in individuals, nor is it fully responsible for their deaths. However, we cannot ignore that an affordance of the bridge is that it facilitates suicide. One San Francisco resident even likened the bridge to a loaded handgun sitting on a coffee table, beckoning self-harm.[42] Obviously, the Suicide Deterrent System will not solve the problem of suicide altogether, but it illustrates how an ethical design recognizes and anticipates the potential for residual effects on vulnerable and under-represented populations and requires an open-minded approach to solutions that prioritize wellbeing.

Designing *for* death

We spend so much time and effort designing our built environment to keep us safe that it would seem counterintuitive to design ways to prepare for death and to die comfortably. In fact, even discussing the topic of death is difficult for many. But since the COVID-19 pandemic, which resulted in millions of deaths worldwide, there has been an increase in the number of "death cafés," or informal group sessions (usually held in café settings) in which people discuss anything death-related. Anyone can download a facilitator's guide and host a Death Café session, and even the National Institute for Health (NIH) recognizes the palliative care of death cafés on "death and grief literacy" of individuals.[43] The topic of death remains largely taboo, despite its inevitability, but some are working to overcome this stigma by advocating for death education and preparedness.

Isabel Knight is a certified "death doula" and the CEO and founder of The Death Designer, a non-profit organization that uses human-centered design to help individuals and families navigate end-of-life "deathcare."[44] I spoke with Knight about how designing for death is ethically driven; she rightly pointed out that "Deathcare is arguably one of the spaces where it is important that we design for the most vulnerable. Because for many people, the thing that happens when they are at the most vulnerable moment in their life is: they die." For Knight, deathcare does not necessarily look the same for everyone. She notes that the most common experience is for people to wait until the last minute to discuss their deathcare plans, and often, families are exploited financially by the capitalist interests of funeral homes. Furthermore, Knight says, "We have a healthcare system that defaults to keeping people alive at all costs, even if it means an extremely compromised quality of life during the last years of life" (Isabel Knight, personal communication, February 15, 2024). For those who wish to die comfortably at home or for those who are unwilling to burden their family members with exorbitant medical bills, the option of "Medical Aid in Dying" (more commonly referred to as *assisted suicide* or *assisted dying*), is neither available nor legal in most places.

As of 2022, physician-assisted suicide is legal in only ten countries and in parts of the United States (11 states, to be exact). In Switzerland, the non-profit organization DIGNITAS is described as "the spearhead for the worldwide implementation of 'the last human right.'"[45] As its name suggests, DIGNITAS offers compassionate end-of-life care for those who wish to "die with dignity." Other countries that permit assisted dying have different regulations under which they must operate. It does seem unethical, however, to legislate whether a person can seek aid in dying and whether others can offer that person deathcare. In Washington state – where assisted dying is legal – a non-profit group, A Place To Die, offers communal spaces to those who wish to exercise their right to die at no cost to them or their loved ones. Their mission is to "actively dismantle systems of oppression as they present in end-of-life."[46] In the book *Tricky Design: The Ethics of Things*, authors Fisher and Gamman argue: "Given that design is employed readily to create the nuances of our *living*...the same opportunities for design should be sought and applied to our inevitable *dying*."[47] Health outcomes have been associated with the deliberate design choices in hospital settings – for instance, paint colors, ambient sounds played in nurseries, and incorporating outdoor green spaces lead to lower levels of anxiety and act as stress reducers. But beyond hospital design, we might also consider how de-criminalizing assisted suicide might benefit highly vulnerable populations. As Isabel Knight explains, "Outside of the interior design choices about the ways that we design our physical spaces like hospitals and nursing homes, we also make design choices when creating policies."

Death-inspired objects also must be evaluated ethically. For example, Tikker is a digital watch that uses basic health profile information and responses to a questionnaire to calculate an individual's life expectancy. The watch then displays a countdown clock to the pre-determined death date. According to the developers, "We have created a product that is aimed at those who really do want to take advantage of every second of their lives." Wearing the Tikker watch would be a constant reminder of impending death and likely take a toll on the user's psychological wellbeing – especially as the death date nears. Another death-inspired object is the Sarco pod, a 3D-printed capsule designed for assisted dying. The mobile "suicide pod" has not yet been approved by any governing bodies, but a prototype shows how it would flood the capsule with nitrogen and slowly asphyxiate the person inside after about ten minutes. Dr. Philip Nitschke, designer of the Sarco pod, has indicated that if the prototype is approved, he will not sell the pod commercially but instead release the blueprints online for anyone to download for free and build their own using a 3D printer. Deathcare should be inclusive and compassionate for all, but it should not serve as the guide for how we live our lives. Whether we are designing against or for death, it is necessary that we are prepared for death but not let it become our life's obsession.

Case study: Emotional robots

The Tamagotchi digital pet was one of the most popular fads of the late 1990s. The handheld egg-shaped keychain had only three buttons and a small digital screen. The game was simple: care for your digital pet by "feeding" it regularly, giving it sufficient attention, and disciplining the pet when it acts out. Neglecting your digital pet for even half a day would result in its death and require a reset with a new egg. Tamagotchi required constant interaction from its users, who were mostly comprised of elementary and middle schoolers, and this eventually led to the digital pet being banned in many school districts in the U.S. because of its distractibility. The term "Tamagotchi craze"[48] referred to the overwhelming desire to keep the virtual pet alive for as many years as possible (one Tamagotchi year = one day in real-time) and the despair of losing the streak and experiencing its digital death. Tragically, in 1997, an 11-year-old girl committed suicide after her parents confiscated her Tamagotchi keychain, and her virtual pet died. Her mother was quoted as saying, "We thought it was just a harmless toy." In fact, the Tamagotchi *was* a toy, and the virtual pet was artificial, but the emotions and attachment were very real.

The concept of "realness" in technology is a highly complex one. When technology replaces human emotional behavior, we question its authenticity. For instance, if you've ever interacted with a chat bot on a website or with an automated voice assistant on a phone call, you probably would not describe that as a "real" interaction as you would a text chat or a phone call with a human person on the other end. It's doubtful that you would consider your relationship with Siri or another virtual assistant on your smartphone anything more than transactional; you ask a question or give a directive, and she complies. Similarly, technology facilitates online romantic relationships through social media, apps, chat rooms, and websites. If you meet someone online and interact with that person solely through the medium of technology, you still might perceive it as a meaningful relationship. But what of the countless number of people who have been "catfished"? (*Catfishing* refers to an online relationship where one person is misled to believe the other person is someone he or she is not). An online relationship can be perceived as real unless it is revealed that it has been an illusion the entire time. This raises the question of whether the expression of love and emotion requires *reciprocity* – or the mutual and equal exchange of feelings – for these feelings to be valid. After all, many of us love our pets, but our love is unrequited. The human-animal bond (HAB) framework explains that humans and animals mutually benefit from one another in terms of well-being but that animals provide humans with emotional support rather than reciprocate that love.[49] Despite this, the *realness* of our human emotions does not depend on reciprocity; just consider the boundless love that you experience for a newborn child, who is not yet mature enough to return that love. So, then, what do we make of technology that is designed to simulate – not reciprocate – human emotion?

Emotional robots are machines that often resemble humans (sometimes called *humanoid robots*) capable of mimicking human emotion. Pepper is one example of a social humanoid "companion robot" developed by United Robotics Group designed to provide compassion, emotion, and conversation with its human partners. It is programmed to interpret human emotions based on facial recognition technology and then assist them accordingly. The website for Pepper succinctly describes its purpose as "Human emotions decoded, embraced, echoed." Among other things, Pepper can be used as a companion for elderly people living alone, playing with young children, and even cheering up hospital patients. It is able to perceive when its human companion is feeling sad or depressed and respond by telling a joke, playing their favorite song, or simply conversing with them. Simply put, emotional robots are designed to fulfill the psychological, social, and emotional needs of humans, but this presents serious ethical issues.

In her book, *Robots and the People Who Love Them: Holding on to Our Humanity in an Age of Social Robots*, Eve Herold warns against the outsourcing of human emotions to machines. She explains that the "socialness" of robots is not authentic, as it only mirrors our emotional energy, creating the illusion of a partner who listens and empathizes. In fact, machines are incapable of such emotion. Their algorithms are programmed by humans with biases, values, ethics, and morals. Pepper, the companion robot, relies on its "emotion engine," which was designed and built by developers to analyze inputs from its human companion and then respond with a pre-programmed action. Data from every interaction are stored in an online database, and this informs future behaviors and responses from Pepper. This immediately raises questions about personal data privacy and its susceptibility to data breaches and malicious hacking. Additionally, technology that learns from a collective database of human behavior is not a good idea, historically. In 2016, Microsoft released its AI chatbot, “Tay,” which was designed to generate responses based on an aggregate of human interactions on Twitter. The chatbot was abandoned less than one day after its launch because of its inflammatory, racist, and generally offensive tweets. Tay did not possess an ethical framework for evaluating its thoughts and statements; instead, it merely parroted back the despicable rhetoric that it had learned from the depths of social media, where anyone can say anything. Technology is deeply intertwined with human behavior, and it is highly speculative to argue that any robot will be capable of thinking or emoting independently of its human inputs. Or, as Herold says, “It may not be the robots we need to fear but human nature itself.”[50]

Emotional robots can be useful. Research has shown that human-centric robots can improve mood, mental health, and even quality of life.[51] They are wonderful as listening devices and for providing basic information and stock responses. You can even stay at the Tokyo Ginza: the first hotel staffed entirely by humanoid robots. But we should be careful not to confuse social and emotional humanoid robots with humans. Nor should we forget the ethical implications that arise from using robots to satisfy our social and emotional needs. As we continue to humanize the machines that surround us, we must consider how we dehumanize ourselves.

Chapter 8

Technology

Can technology be humane?[1]
Paul Goodman

The epigraph for this chapter comes from the author Paul Goodman's 1969 article in the *New York Review of Books.* Goodman was reflecting on the souring of public attitude towards the technology that had been on display during the recent World War: Hitler's eugenics programs, the sophistication of military weaponry, and the scientific breakthrough of nuclear fission represented a pivot from the up close and personal combat of the first World War towards a more technology-based style of warfare that creates more destruction with less human involvement. More than a half-century later, the human soldier is becoming even further removed from death and destruction, with the modern battlefield consisting of cyberattacks, remote-operated drones, and autonomous weapons systems. An *autonomous weapon* is "a weapon system that, once activated, can select and engage targets without further intervention by an operator [person]."[2] Through programmable code and artificial intelligence, an autonomous weapon uses a machine-based algorithm to decide when to use force, how much, and on whom. In other words, we are lessening the dependence on human judgment during times of conflict, when empathy, compassion, and humanitarianism are most needed. Goodman's final assessment of the technological advancements in military prowess was that "technology is a branch of moral philosophy, not of science." As the previous chapters in this book have shown, technological progress will always march forward, so we must ask ourselves not whether we *can* develop the technology, but whether we *should.* This is true of all technology but particularly true of weapons technologies, which are designed for death. And Goodman arrived at an important question: even if a technology is designed with good intentions, can it ever be *humane*?[3]

Every technology has tradeoffs. Marshall McLuhan, the father of new media, wrote exhaustively on the subject. His "tetrad of media effects" is a model that exemplifies how technology creates both extensions and amputations. The tetrad evaluates technology using four questions: (1) What does the medium enhance? (2) What does the medium make obsolete? (3) What does the medium retrieve that had been obsolesced earlier? and (4) What does the medium reverse or flip into when pushed to extremes? The construction of the cross-country highway system, for example, ***enhanced*** travel, making it faster and creating more direct routes to major cities; ***obsolesced*** train travel; ***retrieved*** personal contact and interaction with others across far distances; and ***reversed*** travel times with traffic jams and car pile-ups, and highway construction. Furthermore, there are social and political costs of technology; as I discussed in Chapter 3, highway construction often segregated White and Black communities and trampled sacred lands and historic neighborhoods. There will always be victims of every technology – whether the technology is intentionally or unintentionally misused. By its own nature, technology creates classes: those who can use it and those who cannot, those who benefit and those who are exploited. Because technology is created by humans, and every human is colored by their biases, every technology births a new set of problems, blind spots, and the responsibility to minimize them. As the media theorist Neil Postman wrote, "Technology creates its own imperatives and, at the same time, creates a wide-ranging social system to reinforce its imperatives."[4] The introduction of any new technology not only establishes rules and expectations for its use but also unleashes a ripple effect of behaviors, consequences, and a re-examination of humanity writ large.

In Postman's 1999 book, *Building a Bridge to the Eighteenth Century*, he argues that all innovation should be reduced to a single evaluative question: "What is the problem to which this technology is the solution?"[5] Failed technologies are those that wrongly perceive an existing problem or propose an unnecessary solution to one. Postman uses the example of the Concorde, a supersonic jet that could travel across the Atlantic Ocean in just three hours instead of the usual six. Due to a variety of reasons, the multi-billion-dollar project was scrapped in 2003, and air travel between the two continents resumed its normal flight duration. On a smaller scale, the Segway was a motorized personal scooter launched in 2001, and it was supposed to be a mode of transportation that offered an alternative to either walking or driving a car. City walkers could shorten their commutes by riding the Segway on sidewalks, but this posed obvious dangers to pedestrians and to riders using that same space. In what would become an insurmountable public relations nightmare, the owner of the company died in 2010 after riding his Segway off a cliff in Britain, and the company folded shortly thereafter. It turns out that people are mostly content with intercontinental air travel and with navigating city sidewalks on foot. Both these innovations were trying to solve non-existent problems, and as a result, they failed miserably.

The cardinal sin of modern humans, according to Postman, is that "people believe that technological innovation is synonymous with human progress."[6] But as this book has argued, we should not be developing technology for technology's sake; we should carefully consider the problem that it solves and how it will impact the wellbeing of all who intend to use it and who will be affected peripherally by the technology. I'll return briefly to the original example that I offered in Chapter 1: the K-Cup coffee pod. What problem was the K-Cup solving? Presumably, the pod gave coffee drinkers the convenience of brewing a single cup rather than a whole pot of coffee. However, an ethical assessment of the K-Cup would have predicted that the successful diffusion of K-Cups would lead to a significant amount of plastic waste that would be non-biodegradable. I have developed an ethical evaluation checklist (also included in Chapter 1) to help anticipate these types of ethical problems that stem from new technologies. Similar evaluation tools exist; the Ethical Technology Assessment (eTA) is a checklist tool that includes nine areas of consideration for new technologies: (1) Dissemination and use of information, (2) Control, influence and power, (3) Impact on social contact patterns, (4) Privacy, (5) Sustainability, (6) Human reproduction, (7) Gender, minorities and justice, (8) International relations, and (9) Impact on human values.[7] Failure to meet any of these criteria would trigger a deeper scrutiny of the technology in question and warrant a conversation about the imperatives that it creates. Similarly, the Markkula Center for Applied Ethics at Santa Clara University has developed "ethical pre-mortems," which are questions designed to predict and evaluate the ethical posture of any given technology. These pre-mortems include: Why are we doing this, and for what good *ends*? Will society/the world/our customers really be better off with this tech than without it? Or are we trying to generate inauthentic needs or manufactured desires, simply to justify a new thing to sell? Has the ethical benefit of this technology remained at the center of our work and thinking?[8] Beyond this basic approach to developing ethical things, the researchers at the Markkula Center also advise thinking about new technologies like a criminal would, asking, "Who will want to abuse, steal, misinterpret, hack, destroy, or weaponize what we built?"

The term *technology* has vast implications. Various dictionaries define the term using phrases like "a rational process for creating order," "the use of scientific knowledge to solve practical problems," and "an effort to organize the world."[9] In all definitions, there is an undercurrent of design; usually, technology is developed in response to a specific problem. In this sense, *technology* is a means to an end. Still, this means that an advanced autonomous robot is a technology, but so is a plastic straw. Further, humans can use technology to visit the outer reaches of the solar system, but an elephant can use a broken tree branch to scratch an itch. Technology does not depend on the complexity of the object, or who is using it, just that it is intentionally designed and then applied for a specific purpose. As we consider the examples from this chapter, we ought to remember that many of these technologies are in their infancy and still developing, or they have not yet been adopted by the masses. Their ramifications are still unkown, but it is not too late to take a step back, evaluate them ethically, and make course corrections where necessary.

Virtual realities

Just as we physically design and develop our built environments, we also construct entire virtual worlds to be experienced through a headset and a screen. The concept of *virtual reality* is not necessarily new, but it is becoming far more common thanks to new developments in technology and its increased affordability. At the time of this writing, the Apple Vision Pro has just been released to consumers, and it retails for $3,500. This price point likely will come down over time. Similar headsets like the Meta Quest Pro and the HTC Vive Pro are much more reasonably priced at $500 and $800, respectively. Although here have been many iterations and advancements in the technology, the definition of *virtual reality* has remained relatively unchanged. A 1993 article published in the journal *Virtual Reality Systems*, described VR as "an immersive multi-sensory experience."[10] Fundamentally, this definition is accurate but still somewhat vague. For more, I'll defer to Jaron Lanier, who is considered the father of virtual reality. In his book, *Dawn of the New Everything*, he offers up 52 unique definitions of the term. His first definition reads: "a twenty-first-century art form that will weave together the three great 20th-century arts: cinema, jazz, and programming."[11] His twelfth definition: "VR is the technology of noticing experience itself." His forty-fifth definition: "a person-centered, experiential formulation of digital technology, that hopefully inspires digital economies in which the real people who are the sources of value aren't ignored." His fifty-first definition: "the medium, that can put you in someone else's shoes; hopefully a path to increased empathy." Lanier's rhetorical strategy to define VR in so many different ways illustrates how multi-faceted and diverse virtual reality can be for users. One thing is certain: the multistability of technology ensures that VR will provide us with ample benefits and drawbacks, making it impossible to evaluate whether VR is justifiably "good" or "bad." Although Lanier possesses a healthy skepticism of virtual reality and its promises, he believes it is "the most humanistic approach to information." This is because the fully immersive experience strips away the gadgets and devices and leaves the user with just the individual experience, fooling them into a falsely constructed reality, but one that "exposes you to yourself" and "proves you are real." In a seemingly contradictory way, Lanier's view is that our virtual selves reinforce our real selves.

As I mentioned in the previous chapter, there is a paradox that exists between technology and realness that depends on your definition of what it means for something to be "real." If perception is reality, then VR is real. If reality only exists in the shared, observable, lived experiences between physical beings, then VR is only a simulation of reality. Regardless of our varying interpretations of reality, we can hopefully agree that the virtual reality environment – as with all technology – has been created by humans, for humans. Therefore, as we devote more of our time to the virtual, it is imperative that we not simply debate whether VR is real or not, or whether it is good or bad for humanity, but instead, evaluate the ethical challenges of the virtual infrastructure.

Charles Seife's groundbreaking work, *Virtual Unreality*, positions VR as a potentially harmful playground. He notes that the internet has already altered the way in which we interact with our surrounding world. Just look around any café, classroom, or public park (or really any public setting, for that matter) to observe how we cling to our devices for a variety of reasons: to assuage boredom, to avoid social awkwardness, to maintain constant connections, and more.[12] But Seife points out a key observation: in the virtual world, "people are using the properties of digital information to try to alter your perception of reality."[13] It is not enough to examine how we use digital technology; we also must address how it shapes our realities. The ubiquity of the internet and the subsequent rise of social media fuel the extreme digital polarization that we experience online, which then translates to in-person experiences. The social and political unrest in the United States throughout 2020 and 2021 exemplify how online movements foment real-world clashes. For example, the Black Lives Matter and Blue Lives Matter movements are just two examples that are politically-opposite of one another but that share the same origin story: they were both conceived online. (Please note that I am not equating the two movements; Black Lives Matter clearly advocates for positive social change, and Blue Lives Matter is an intellectually dishonest, tone-deaf, and unrelated counter argument for the support of law enforcement). The point is this:

regardless of their political and social motives, these examples both illustrate how the voices of online extremists drown out more moderate voices and how this leads to divisiveness that sometimes translates into real-world actions such as violent protests. In Seife's words, "The digital revolution has dramatically changed not just how we gather information about the world, but also how we can tamper with the information that others are gathering."[14] We receive our news and information through self-selected filters that reinforce our beliefs and further separate us from the opposing viewpoint. This is true of our current digital landscape, but VR hopes to immerse its users even further into this virtual existence. Seife concludes, "We have become reality engineers."

As I have referenced throughout this book, there are ethical standards and codes that nearly every discipline establishes. Physicians, engineers, architects, realtors, and lawyers are just a few examples of professionals who are held accountable to their industry's code of ethics. The American Contract Bridge League (ACBL) is the largest association of recreational bridge players in North America, and its 165,000 registered members are expected to uphold their organization's ethics policies while playing games. Even sanitation workers – whose work is largely under-appreciated – agree to the ethical code articulated by the American Board of Industrial Hygiene in all matters of waste management. It is a reasonable assumption, then, that we expect designers and developers of technology and virtual reality environments to be held to ethical standards to ensure the wellbeing of users and of our society in general. But, anticipating ethical risk is more difficult when dealing with innovative technologies than it is for other disciplines. For instance, the Code of Ethics for the American Institute of Architects includes the section "Human Dignity and the Health, Safety, and Welfare of the Public" (E.S. 1.5), which states that "Members should employ their professional knowledge and skill to design buildings and spaces that will enhance and facilitate human dignity and the health, safety, and welfare of the individual and the public." Ostensibly, it would be unethical to knowingly build a pedestrian foot bridge that does not meet rigorous engineering standards. But requirements such as the capacity, span, structural load, and even volume of materials are scientifically calculated; we know the bridge will fail if it does not meet these specifications. With novel technologies, such as highly realistic VR environments, we might not foresee what types of risk it will pose or even whether it will be adopted. After all, if I had said to someone fifteen years ago that Instagram poses a substantial risk to the emotional wellbeing of young teenage girls, the response would have been, "What is *Instagram*?" Technology moves fast. It is much easier to look in the rear-view mirror and evaluate a technology than it is to predict its ethical positivity.

There are promising uses of virtual reality. For instance, virtual reality-based therapy, or VRT, can increase accessibility to licensed therapists. A sub-genre of VRT is virtual reality exposure therapy (VRET), in which individuals are immersed fully into a virtual environment to work through an array of anxieties and phobias. For instance, someone with a fear of heights can confront their phobia virtually, using exposure therapy to address their fear while minimizing the risk of physical harm. VRET has even been shown to reduce bodily pain; a 2022 study published in *Annals of Behavioral Medicine* found that burn victims experienced a 35-50% reduction in physical pain when they were immersed virtually into a cold environment.[15] Future VR applications might include the ability to reconnect with loved ones after they have passed on. Jang Jing-sun was the subject of a South Korean documentary in which Jing-sun was reunited with her deceased daughter using virtual reality. Having lost her 7-year-old daughter to blood cancer in 2016, the documentary *Meeting You* reimagined the young girl virtually so that her mother could reunite with her once more. Similarly, educational applications of VR might one day allow students to interact with historical figures and experience historical events firsthand. Obviously, there are serious ethical implications for exposing people to hyper-realistic situations, particularly if it leads to a new or relived trauma.

In his book *The Ethics of Virtual and Augmented Reality*, Erick Jose Ramirez proposes The Equivalence Principle (TEP). According to Ramirez, "If it would be wrong to subject a person to an experience, then it would be wrong to subject a person to a virtually real analog of that experience."[16] The authors of an article published in *Frontiers in Virtual Reality* note that even though VR is not yet technically capable of creating an environment that can be mistaken for the real world (this is referred to as *superrealism* and certainly is a possibility), it is nevertheless "real as an experience." In other words, even though VR users understand that they are in a virtual environment, they behave in real-world ways and experience real-world effects from VR experiences. They add, "We change through our experiences: experiences produce changes in the body and the brain. In other words, just as real-life experiences have after-effects, so virtual experiences may have physical, emotional, and cognitive after-effects which may be beneficial or harmful."[17]

VR may only be a simulation of reality, but its effects are very real. For instance, there is potential for significant physical harm as a result of using VR. Chances are that you have seen online videos of people running into walls, falling down, and being generally disoriented while using a VR headset (in fact, these videos are regularly featured on *America's Funniest Home Videos*). While most people in these videos suffer only from a bruised ego, it can be far more dangerous. In 2017, the first reported VR-related death stemmed from a Russian man whose impaired vision from his VR headset caused him to fall through a glass coffee table and die from excessive blood loss. Although this is not common, over-exposure to VR can have devastating psychological consequences, including but not limited to, harmful exposure and desensitization to sensitive topics and mature content like violence and pornography, the intensification of feelings that make it difficult to leave the virtual world, and the desire to prioritize life in the virtual over the real world.

In 2022, the founder of one of the first VR headsets, the Oculus, launched a prototype for NerveGear: a VR headset capable of killing its user. When the user's character dies in the virtual reality game, Sword Art Online, the headset detonates three – very real – explosive charges mounted to the user's head, killing them instantly. The developer, Palmer Luckey, wrote in his blog: "The idea of tying your real life to your virtual avatar has always fascinated me – you instantly raise the stakes to the maximum level and force people to fundamentally rethink how they interact with the virtual world." Luckey describes the NerveGear as a "piece of office art" that serves as a "thought-provoking reminder of unexplored avenues in game design." Though he adds that, "It won't be the last."[18]

Virtual environments pose serious ethical implications that must be addressed. For example, how do we ensure user data and privacy are protected? Who is accountable for physical or psychological harm that occurs in virtual reality? Should virtual reality applications be designed as a superrealistic form of digital escapism that excuses immoral behaviors, or should they be designed to promote wellbeing? In all likelihood, virtual reality will be designed to offer both. In the coming years, we will witness a mass exodus from the real to the virtual, just as human interfacing has shifted from the person to the screen. It is critical that we remember who has designed this virtual reality, their motivations for doing so, and how we might be exploited in the process.

Surveillance

Your image is captured on surveillance video an average of 70 times per day. Now, this depends heavily on where you live (urban or rural setting) and how often you travel (commercially or personally), and both variables can substantially increase or decrease this number. Regardless, once you leave the confines of your own home, you can almost certainly plan on being recorded and/or photographed in public spaces. In addition to the CCTV security cameras that stake out the exterior and interior of businesses and the traffic cameras that litter roadways and intersections, there is a growing number of residential homes that utilize a video doorbell camera. In 2021 alone, there were 12 million video doorbells sold globally, about 30% of which were Amazon Ring video doorbells.[19] If you have ever watched an episode of the true crime series, *Dateline*, you would know that video surveillance is often the first step of the investigation into someone's whereabouts. By piecing together footage from residential homes, businesses, and roadways, detectives can often verify (or debunk) a suspect's timeline, and this is considered ironclad evidence. Then, there is the high likelihood that you have appeared in the background of someone else's photograph. In Rose Eveleth's piece for *The Atlantic*, titled "How Many Photographs of You Are Out There in the World?," she contemplates the number of photos that are taken and published to social media during any given day and then asks, "How many photographs out there feature you or me in them, in the background, beneath the Eiffel Tower or at the table next to the birthday party? Is it 100? 1,000? 5,000? 10,000? More?" Again, this number is highly dependent on your location, but it illustrates how we are unknowingly captured in photographs that end up being circulated online.

Hannah Fry's book, *Hello World: Being Human in the Age of Algorithms*, aptly describes our tricky relationship with surveillance. On the one hand, constant surveillance can provide safety – we are less likely to behave badly when we know we are being recorded – and accountability for those who do wrong – a criminal is more likely to be caught and prosecuted successfully if the crime is caught on camera. But on the other hand, relinquishing control to technology unlocks new sets of problems. An example of this is the use of facial recognition technology by law enforcement to identify criminals. A 2022 Pew Research study found that most Americans see facial recognition technology as a beneficial tool for law enforcement agencies to use in solving crimes, though younger people are more highly skeptical of police surveillance.[20] Fry points out that the primary risk of facial recognition technology is misidentification, and that this probability is considerably high; even the best facial recognition software has only about an 80% identification rate accuracy. And facial recognition technology disproportionately misidentifies women and people of color.

The Gender Shades project was a groundbreaking initiative at the MIT Media Lab that ran from 2017 to 2020. Its main purpose was to evaluate the leading technologies in facial recognition for their predictive accuracy. The project's researchers found that in all cases, facial recognition software was more accurate in predicting the identities of light-skinned males but performed the worst in accurately identifying dark-skinned females. This finding replicated earlier studies, which had suggested an algorithmic bias in the functionality of the software. There are multiple factors that contribute to this bias; images of dark-skinned people have fewer features to contrast, the use of cosmetics on women can disguise natural features, and perhaps most importantly, facial recognition technologies perform best on the demographics on which it was trained. The Center on Privacy & Technology at Georgetown Law states that "training is destiny; the faces that an algorithm practices on are the faces it will be best at recognizing."[21] They add that most facial recognition technologies are trained on models of white males and that designing a system to perform well on one demographic often comes at the expense of the others. To put it more succinctly, our technologies – and the algorithms upon which they are built – are not neutral. They reflect the perspective of the developer, who, in the field of computer science, is predominantly white male.

Even if we could design a facial recognition tool that had a 100% accuracy rate, this still might not justify its use ethically. The researchers of the Gender Shades project warn that "flawless facial analysis technology can be abused in the hands of authoritarian governments, personal adversaries, and predatory companies."[22] Consider a hostile government that uses facial recognition technology to identify and then apprehend those who are critical of authority or who are participating in the protest, or a social media platform that automatically identifies and tags the people in photos, or a retail store that knows when you are near and can spam your phone and email with offers. Fry wants us to think of surveillance technologies not as objective, non-neutral tools, but as a "source of power."[23] The 2022 Pew Research study also asked respondents how they think law enforcement will use facial recognition technology; most admitted that it would be used to surveil Black and Hispanic neighborhoods at a disproportionate rate.

Some advocacy groups are campaigning for awareness of how and where facial recognition technology is being used. For example, the Electronic Frontier Foundation (EFF) hosts the website whohasyourface.org, where users can take a 5-item quiz to determine which state and federal agencies have access to a database containing your photo that can be used for facial recognition. The EFF is working to ban the use of facial recognition technology by law enforcement, citing the obvious flaws in the technology and the unethical treatment of innocent people. The organization likens the use of facial recognition technology to being in a "perpetual lineup" without even knowing. But while some are working to adjudicate the unfair use of facial recognition, other countries are choosing to embrace the surveillance state.

In 2014, China announced its Social Credit Plan – a national initiative to boost surveillance on its citizens and use this data to create a digital profile that would inform nearly all aspects of individuals' lives, including medical, legal, social, educational, transactional, and even recreational activities and opportunities. The aim of this plan is to bolster economic value and foster a more secure public by implementing two scoring systems: financial creditworthiness and social creditworthiness. The first system (*zhengxin*) is not that different from the Western model of using a consumer's history of transactions and financial activity to rate them – much like a FICO credit score does. But the second system, social creditworthiness, is far more interesting. In this system, the Chinese government is looking to build trust both between government institutions and individuals, and among individuals. To do so, behaviors and activities are monitored and then scored; in the same way that financial behavior dictates financial opportunity, social behavior can impact social image and mobility. Chinese government officials are still exploring ways to implement the social creditworthiness system, but some local municipalities have begun experimenting.

In Rongcheng, a city of roughly half a million people, residents are assigned a point value according to their good and bad actions. According to the Brookings Institute, each Rongcheng resident starts with 1,000 points, and every good and bad action either adds to or subtracts from this total. For instance, donating blood would reward points, but receiving a traffic ticket would deduct points. Similar to how a low credit score restricts the ability to secure a loan or imposes higher fees on someone deemed "at risk," a low social credit score would restrict individuals from buying a home or accessing high-speed internet because they are "untrustworthy." The Center on China's Economy and Institutions at Stanford University reports that Chinese citizens are largely in favor of this social credit system, though the majority of support comes from those with higher socioeconomic status and who possess strong faith in the Chinese government. Specifically, there is strong support for surveillance of behaviors in the following areas: conduct on public transportation, blood donation, spreading rumors online, domestic violence, protesting the government, and even sorting household trash. Chinese citizens indicated support for social credit system punishments, including restrictions on job promotion, travel, internet use, and children's admittance into private schools; disqualification from civil servant jobs; and publishing social credit scores publicly.[24] It is important to note that the Rongcheng approach is experimental, and the Chinese government has not indicated that this will be the model used for the national social credit system, but there is an important lesson to be learned here: people are generally supportive of hyper-surveillance when they feel that it benefits them, making them less sensitive to the ethical tradeoffs that might impact those who are more vulnerable.

It's not hard to imagine a Western city that adopts the Rongcheng model of surveillance. In fact, we're nearly there. Your everyday movements and behaviors are captured on a plethora of cameras, yet we hardly think to ask who owns that data and what they are doing with it. Questions of transparency, accountability, and data privacy are deeply embedded into the ethical design choice to use surveillance technology. Plus, we might question whether heightened surveillance even works. The company SmartSecurity Pros offers a wide range of surveillance technology for K-12 schools. According to its website, school surveillance creates a more peaceful school climate, where "Bullies, drug dealers, and other rule-breakers are forced to think twice."[25] But, empirical research suggests otherwise. Longitudinal data on "high-surveillance schools" clearly shows that an increase in security measures such as metal detectors and AI-based security cameras corresponds to more disciplinary actions over time. More surveillance yields more offenders. There are residual effects of hyper-surveillance, as well. Students attending high-surveillance schools have reported feeling *less* safe and wrongly perceiving *more* violence in their schools. Given what we know about the mindset that hyper-surveillance creates, the algorithmic biases that are baked into its design, and the ethical tradeoffs that surveillance presents, we might ask ourselves whether it is worth it.

Let's apply Paul Goodman's question from the start of this chapter to the issue of surveillance: can surveillance technology be humane? I suppose the answer depends heavily on your perspective. If you are a law enforcement agency that has used facial recognition to successfully identify and capture a killer, then surveillance is humane. But during that same process, numerous people might have been misidentified and wrongly perceived as being criminals simply because their image and likeness exist in a database that acts as a perpetual virtual lineup. The question is not whether constant surveillance is a humane tool, but rather, to what extent we are willing to trade privacy for security. The Center for Humane Technology outlines six tenets that evaluate whether a technology is "humane": (1) Sensitive to human nature, (2) Minimizes unintended consequences, (3) Centered on values, (4) Narrows the gap, (5)

Builds shared understanding, and (6) Helps people thrive.[26] I contend that surveillance technologies are not sensitive to human nature, that they create unintended consequences, and that they stifle rather than help people thrive. This is not to say that there is no place for surveillance: of course there is. But we should not count on surveillance (or any technology, for that matter) to preserve *humanity*. To do so is counterintuitive; technology can be designed to simulate compassion, but it is not compassionate. As I discussed in Chapter 5, Jeremy Bentham understood how the threat of surveillance changes people's behaviors. As we mount more and more cameras to our front doors, our schools, storefronts, and even public spaces, we have created our own digital panopticon in which we behave as if we are being watched (or, at the very least, with the understanding that anyone can pull out their smartphone at a moment's notice and record an event). Yes, I suppose it is worthwhile to wonder whether this is a humane environment that we've created for ourselves.

Autonomous vehicles

Our *built environment* includes every aspect of our lived experiences: the fixed architecture and infrastructure that occupies our space, as well as the products, processes, and systems that mediate our behaviors within this space. And while the built environment is continually evolving, technological progress usually means outsourcing human talents to machines that can simulate that effort, often more efficiently. This is especially true of the evolution of transportation. Imagine an early man needing to cross a strait of water; assuming he could swim, his only option to reach the other side would have been to swim across the strait, that is, until the concept of a floating vessel was realized. At this point, you wouldn't have needed to know how to swim necessarily, but you would need to exert physical effort to row, paddle, or push the vessel. This brilliant invention replaced a great deal of human effort and likely birthed entire seafaring cultures. Crossing bodies of water would have gotten even easier with the introduction of a cloth sail that could harness the wind. Now, greater nautical distances could be traveled, though there was an even further shift away from human power and towards a reliance on natural wind energy. Next, steam-powered engines were developed to propel ships more efficiently and reliably, followed by gas turbine engines, which paved the way for modern fleets of nuclear-powered vessels that depend on sophisticated computer software systems that are often operated from a central command center using joysticks and computer code. This offloading of human effort onto machines is a recurring theme that runs throughout the entire human record and has become synonymous with technological progress.

As we evolve, we outsource more of ourselves to machine automation, but this is not exclusively a bad thing. Living in a modernized first-world country affords me certain luxuries that bypass human effort. For example, I have access to clean drinking water at the turn of a faucet handle rather than having to scoop it out of a river; I can summon lighting at the flick of a switch rather than having to fuel candles and torches; I can achieve precise heating and cooling temperatures by pressing a few buttons on my thermostat rather than tending a fire or sweating through hot summer nights. This automation is not magical; it is the scaffolding of layers of design, experimentation, refinement, and redevelopment of ideas, prototypes, and models that began long before we were here. Every technology has a historical precedent upon which it is built. And to me, it seems that the evolution of technologies aims to lessen human effort while preserving humans as the sole beneficiary. But as we know, technology simultaneously unlocks new potential *and* creates new problems.

Nicholas Carr is a well-known technology skeptic who has written numerous books and articles on the human consequences of technology. His 2008 article in *The Atlantic*, "Is Google Making Us Stupid?" was widely circulated and remains highly criticized, and his 2010 book, *The Shallows* (which I have referenced earlier in the book), was a Pulitzer Prize finalist. Carr's forewarnings of technology and its impacts run deep throughout his work and have been largely validated as time passes. For this reason, I pay special attention to Carr's writing, which seems to consistently identify the ways in which burgeoning technologies make us less human. His two most recent books, *Utopia Is Creepy And Other Provocations* and *The Glass Cage: How Our Computers Are Changing Us*, examine the impacts of computer automation. Throughout these books, he often returns to the story of the autonomous car (sometimes referred to as a self-driving car) to illustrate how automation can be designed to replicate the menial, rote tasks that humans perform but often falls short when we ask it to perform *intelligently*, or humanly.

There are many actions that are already automated in the cars we drive. Power windows, power seats, power steering, automatic braking, assisted parking, and even the car radio, which scans the airwaves for a clear radio signal, are only a handful of examples of how we have assigned basic car-driving tasks to be performed by computer automation rather than by human effort. And most of the time, this is not an issue. The promise of automation is that it assumes our psychomotor skills and tasks so that we can free up our mental effort and reallocate it to more important things like watching for pedestrians crossing the street, observing construction zones, following posted speed limits, and avoiding reckless drivers. Driving a car requires more than just the physical ability to operate a vehicle; it requires a heightened sense of awareness and responsiveness to unpredictable situations. This, by the way, is why we outlaw driving while impaired and driving while distracted. A self-driving car must be more than a computer program that automates driving skills; it must perform the same cognitive tasks required of human drivers.

The term *autonomous vehicle* has layered meanings. The Society of Automotive Engineers (SAE) has defined six levels of vehicle autonomy, ranging from Level 0, which is the total manual control of the vehicle, to Level 5, which is the fully autonomous operation of the vehicle with no human involvement. At Level 5, the human is merely a passenger in the vehicle, whereas all other levels have varying degrees of human control. Many popular automotive brands like Tesla, Volvo, Audi, GMC, and others offer semi-autonomous vehicles that include a "hands-free" driving feature that relies on environmental detection capabilities like geofencing and LIDAR to sense their surroundings, but these all require constant vigilance from a human operator who can override the system at any time (Level 4). The Insurance Institute for Highway Safety predicts that there will be more than 4.5 million semi-autonomous vehicles on U.S. roadways by the year 2040, and tech giants like Amazon and Google have been testing fully autonomous vehicles (Level 5) on public roadways since 2009. The Self-Driving Car Project – originating from Google and now known as Waymo – has driven more than 20 million miles across 25 different U.S. cities. However, the public attitude towards driverless cars remains distrustful. A 2022 Pew Research Center study found that most Americans would not want to ride in a fully autonomous driverless car, do not feel comfortable sharing the road with them, and feel that driverless cars are a generally bad idea for society.

There is a convincing argument to be made that driverless cars could, one day, reduce the number of vehicle deaths. In fact, the National Highway Traffic Safety Administration (NHTSA) reports that there are 1.25 million vehicle-related fatalities each year worldwide, 94% of which are caused by human error. If we remove the error in human judgment, driverless cars could save countless lives. But an article in the *National Library of Medicine* points out that there are many different strategies that could reduce vehicle fatalities, including improved infrastructure design, better driver education and training, and stronger community engagement and partnerships with law enforcement.[27] In fact, the idea that driverless cars can save us from ourselves is a sort of misguided techno-solutionism. In 2016, the first fatality involving a semi-autonomous vehicle, the Tesla Motors Model S, resulted from a system failure to recognize an oncoming tractor-trailer. An investigation by *The Washington Post* revealed that since 2019, there have been 736 crashes and 17 deaths involving Tesla's Autopilot (a Level 2 semi-autonomous, driver-assisted feature). In 2018, Elaine Herzberg was the first person to be struck and killed by a fully autonomous vehicle. The Volvo SUV was part of a flagship program by Uber to use driverless car technology to shuttle passengers to their destinations. As a preventive measure, each vehicle also employed an Uber chauffeur whose job was to monitor driving conditions and intervene in an emergency. Sadly, Herzberg was not identified by the vehicle's sensors, and the Uber chauffeur, Rafaela Vasquez, was distracted by her phone. Vasquez was charged with negligent homicide before pleading guilty to the lesser crime of endangerment and receiving a three-year supervised probation. And then, in 2019, the first felony case of manslaughter was brought forth involving a crash where the motorist, who was relying on a partially automated driving system, ran a red light and killed two other drivers. Charges are still pending.

Carr emphasizes how an over-reliance on intellectual technologies creates a degeneration effect among its users, who trade their mental acuity for automation (this might be summed up with the phrase "use it or lose it," or as Carr writes succinctly, "Sharp tools, dull minds"). When we outsource the psychomotor tasks that are required for operating a vehicle, we also become cognitively absent – a condition that researchers call *automation complacency*. This is similar to what Langdon Winner described as *technological somnambulism*, which I discussed earlier in Chapter 3, and is a primary concern among technology critics. It's not that skeptics like Winner and Carr are anti-technology; they are pro-humanity. In *The Glass Cage*, Carr states:

> The choices we make, or fail to make, about which tasks we hand off to computers and which we keep for ourselves are not just practical or economic choices. They're ethical choices. They shape the substance of our lives and the place we make for ourselves in the world. Automation confronts us with the most important question of all: What does human being mean?[28]

So, while computers can perform many tasks better and more efficiently than humans, we must contemplate deeper, ethical questions about whether that technology ought to.

Brian Merchant's book, *Blood in the Machine,* gives a detailed account of the Luddite movement in nineteenth-century England. A *Luddite* is a term that we use today to describe anyone who is less than savvy with technology or who is slow to adopt an emerging technology. Someone who uses a flip phone, who pays with cash instead of a card or smartphone, or who remains absent from social media, might be called a Luddite. Perhaps one day, "Luddite" will be used to describe someone who prefers to drive their car manually rather than be shuttled around by driverless car technology. The Luddites have become synonymous with rebellion against technology because this group of workers resisted the automation of their highly skilled labor in the textile industry. The Luddites protested peacefully, and they burned down textile factories with modernized machinery. Their streak of industrial "machine-breaking" has been mischaracterized as being anti-technology, anti-progress. But as Merchant points out, "[The Luddites'] problem lay not with the machines themselves, but with the men who owned them." It's not that this loosely organized group of workers opposed the looming machinery; they felt unfairly treated and exploited by factory owners who valued profit over people. Automation, at its core, is not necessarily about advancements in technological progress and our eagerness to adopt them. According to Merchant, "Automation is, quite often and quite simply, a matter of the executive classes locating new ways to enrich themselves, not unlike the factory bosses of the Luddite days."[29]

The autonomous vehicle market is projected to reach $300 to $400 billion in shared revenue by 2035.[30] Unfortunate stories like that of Elaine Hernandez and Rafaela Vasquez remind us of the human cost of self-driving cars. Research on fully autonomous vehicles is somewhat vague, but early findings suggest that self-driving cars reduce the number of fatalities and police interactions compared to human drivers. But, this doesn't tell the whole story. Autonomous technology is not perfect, and there certainly will be fatalities that could have been avoided if a human driver were at the helm. So, it would seem building an environment where human drivers are a thing of the past is an ethical choice, not a technological one.

We must remember that autonomous vehicles not only relinquish our power to drive but then transfer that power to the company that designed and manufactured that car. Technology should make our lives easier and more equitable, but it shouldn't abdicate humanness. As we allow more of these convenience technologies to direct our lives, we become more complacent. We expect more from technology than we do from each other. The leading autonomous vehicle manufacturer, Waymo, promises to make "frustrations and concerns with driving a thing of the past."[31] But these difficulties are the very things that form the callousness of the human experience. We are human because of the challenges we face, not despite them.

Case study: Killer drones

Isaac Asimov is perhaps the best-known science fiction writer. Although he did not invent the killer robot genre, he perfected it throughout his prolific storytelling career, which began in 1940 and includes more than 380 short stories. As with most sci-fi thrillers, the tension in Asimov's stories hinged on the conflict between man and machine. The formula goes something like this: shortly after an advanced computing or technology has been introduced into the world, it reaches a heightened consciousness that eventually surpasses its human captors, unleashing a torrent of destruction on humanity. The narrative formula continues to work; an IMDB search for movies tagged with the keyword "killer robot" yields more than 250 titles. In the past two years, movies like *M3GAN* (which features a killer doll) and *Five Nights at Freddy's* (which features murderous restaurant animatronics) have earned $181 million and $132 million worldwide, respectively. People are simultaneously enthralled with technology and deeply fear its unpredictable outcomes.

The title of this case study, "Killer Drones," sounds like the title of an Asimov dystopic short story, but it is very real. "Killer Drones" is more of a pejorative; the preferred term by the United States military community is Remotely Piloted Aircraft (RPA), which sounds much less frightening. An RPA is an autonomous weapon – a subcategory of "unmanned aerial vehicles" (UAVs) – that comes in various sizes ranging from 55 to 130 lbs. The lightweight aircraft can be flown from anywhere under the control of GPS and satellite communications and has the capability to strike a target from as many as seven thousand miles away. Advocates for RPAs will recite two talking points, specifically: (1) that this attack method is more accurate and kills fewer civilians than traditional bombing methods, and (2) that RPAs are more cost-efficient than traditional air strikes. But both these claims have been widely disputed. First, the accuracy of RPAs might be improved, but this results in an increase in demand for the number of drone strikes, which leaves more "collateral damage" in its wake. And second, this increase in drone strikes means that their cost-efficiency is not necessarily a cheaper solution than conventional bombing; additionally, the technological development of American military drones has far exceeded original estimates.

So, while RPAs have both benefits and drawbacks, one thing is for certain: they are ethically questionable as weapons of war.

An article published in *The Independent Review* describes the "Drone Paradox," which states that the use of drones in robotic warfare creates more enemy combatants than it eliminates. According to the authors, "The U.S. government justifies its use of drones as an efficient method for weakening and ultimately ending the threat of international terrorism while protecting members of the American military. The use of drones, however, creates and perpetuates terror among foreign populations."[32] Fundamentally, RPAs likely create more of the problem they are trying to solve. The article authors go on to explain a number of other issues related to the drone program, citing a lack of transparency from the government, potential violation of international laws, the precedent that it establishes for other countries to use robotic warfare, and the psychological effects that it has on drone pilots.

Michael Phelps's book, *On Killing Remotely: The Psychology of Killing with Drones*, details the psychological and emotional costs of killing from afar. He compares drones to other advancements in technology that afford killing over long distances. The rifle, for example, created space between combatants. Killing no longer needed to be intimate, hand-to-hand combat; it could be done remotely. This, according to Phelps, makes the act of killing easier because it allows humans to "overcome the built-in mental governor" that is the natural aversion to taking another human's life.[33] Creating distance between combatants effectively dehumanizes the other person – making him a target, not a human – and eases the resistance to kill. This is exacerbated even further when the target is not within the scope of a rifle but instead a virtual representation on a screen. Deploying an RPA on a target creates a vast emotional distance between the killer and the killed, granting moral permission to perform an immoral act. Moreover, remote drone pilots are commanded to use specific rhetoric when executing their strike missions. Phelps points out that the words *kill* and *person* are rarely heard in the context of tactical communications. A "MAM" (military-aged male) refers to an adult male; "target" refers to the person or persons selected to receive the attack; "squirter" refers to a person who survives and then flees such an attack. Further, the linguistic dehumanization of the enemy extends to racial epithets that I do not need to repeat here. "Othering" the enemy becomes easier when they are a blip on the screen.

Aside from the empathetic detachment that comes with killing remotely, there are other ethical concerns with using RPAs. For instance, the human operator could be acting on misinformation or bad intelligence about the intended target, or they could simply be overworked and tired, leading to clerical mistakes that result in civilian deaths. To reduce this likelihood, security experts believe the future of warfare will feature RPAs that have been programmed with artificial intelligence. The AI-guided drones will seek combatants based on an algorithmic dataset and then destroy them. We are already witnessing this shift towards robotic warfare in the conflict between Russia and Ukraine, which has been called "the first full-scale drone war."[34] But as we relinquish more control to our technologies, it is worth asking whether the use of RPAs is ethically justifiable.

In his fictional work, *Handbook of Robotics, 56th Edition, 2058 A.D.,* Isaac Asimov penned the Three Laws of Robotics outlined below.

> **The First Law**: A robot may not injure a human being or, through inaction, allow a human being to come to harm.
>
> **The Second Law**: A robot must obey the orders given to it by human beings except where such orders would conflict with the First Law.
>
> **The Third Law**: A robot must protect its own existence as long as such protection does not conflict with the First or Second Law.

Later, Asimov added the "Zeroth Law," which was meant to inform the other laws. It reads: "A robot may not harm humanity, or, by inaction, allow humanity to come to harm." Together, these four laws seem like an ethically positive framework for co-existing in a technological world. But they are obviously not applied in our real world. Or, as roboticist Daniel Wilson puts it, "Asimov's rules are neat, but they are also bullshit."[35] For example, killer drones are designed to violate each of these laws. But as Asimov would eventually point out, there is a loophole. The term "human" is not clearly defined; if a group of people is considered less than human or even non-human, then they would not be protected.

Chapter 9

Designing the future

The world isn't broken. It's working exactly as it was designed to work. And we're the ones who designed it.[1]
Mike Monteiro

Future generations will experience the world differently from us, just as we experienced it differently from our parents, and they from theirs. The designers, engineers, builders, and consumers of the future are watching us, and learning from our behaviors. And change is afoot. In her 2019 speech to the United Nations, 16-year-old Greta Thunberg chastised her audience for failing to act on climate change. To the world leaders assembled in the room, the teary-eyed Thunberg said, "You are failing us... The eyes of all future generations are upon you."[2] Demanding action from the UN, Thunberg added, "We will not let you get away with this." Her fiery speech garnered plenty of criticism from adults. Then-president Donald Trump posted sarcastically online that Thunberg was "a very happy young girl looking forward to a bright and wonderful future." *Fox News* pundit Michael Knowles called the 16-year-old "mentally ill", and Thunberg has been called a "millenarian weirdo," a member of the Hitler Youth, and "hysterical," among many other dismissive insults. But the polarizing young activist has been successful in expressing the shared helplessness of her generation.

It is my assertion that while it would be ideal to gain a consensus that climate action is necessary, the toxic discourse suggests otherwise. A 2023 Pew Research Center study showed that more than one-quarter of Americans do not think that climate change is a serious problem, and another 13% feel that individuals cannot effect change.[3] The general sentiment – while sharply divided along political party lines – seems to suggest that individuals are not in agreement that climate change requires action, or for that matter, even exists, and if it is recognized, they feel there is little that they can do. One way to overcome this is with ethical design practices that guide users towards positive behaviors rather than relying on them to make an ethical choice. For this, we need to include the younger generations. In the UK, for example, the Royal Academy of Engineering hosts the Teen Tethics Initiative, which strives to connect teenage innovators with professional engineers to provide ethical solutions to case studies that exhibit engineering problems facing difficult decision-making. The program seeks to engage youth voices in shaping future cities – which they will one day inherit.

The Global Youth Outlook is a report funded by an international accounting and advisory firm, PricewaterhouseCoopers, that surveyed 700 individuals aged 10 to 20 years old across 81 different cities across the world. Their project sought to understand youths' outlooks on the future as they pertain to the 17 Sustainable Development Goals (SDGs) established by the 2030 Agenda for Sustainable Development that was adopted by all 153 member states of the UN in 2015. Respondents were asked which of the SDGs was most important to them; the top two most prioritized goals were *Goal 6: Clean Water and Sanitation* and *Goal 13: Climate Action.*[4] Not surprisingly, the climate is an essential focus for younger generations. But the least prioritized goal, according to these survey respondents, was *Goal 9: Industry, Innovation, and Infrastructure.* This includes the targets of creating sustainable infrastructure for developing countries, and developing sustainable, resilient, and inclusive infrastructures. I contend that these goals are inextricably tied; one of the main pathways to climate reform is through our industries, innovations, and infrastructures. We must encourage others to see design as an agent of change – both through the resistance of unethical designs and the embracing of designs for wellbeing.

My undergraduate students in New Media and Design tell me that I can be overly skeptical at times. I don't disagree. I mean, the previous eight chapters highlight many unethical designs or, at the very least, raise questions about their ethics. But I don't consider myself to be a negative person or a pessimist. I consider myself a student of the schools of Carr, Winner, Postman, Norman, and McLuhan, who all argue that design informs the way that we perceive and understand the world. Design is responsible for more than just technological progress; it is an expression of our humanity, and sometimes, I am disappointed by this. So, I'm working on refocusing my attention on what is good – not what is wrong with the world – and concentrating on solutions, not problems. The epigraph for this chapter comes from Mike Monteiro, an acclaimed critical designer and noted cynic of design. I left off the latter half of the quote, mainly for its vulgarity, but I feel it's necessary now to include the rest of it. In his book, *Ruined by Design,* Monteiro says, "The world isn't broken. It's working exactly as it was designed to work. And we're the ones who designed it. Which means we fucked up." I thought the quote was especially poignant because it answers the quote that I selected as the epigraph for Chapter 1. There, Langdon Winner asked, "Just as we 'make things work,' what kind of world are we making?"[5] I think about this quote often, and I try to remind myself that just as there are many examples of bad design, there are many hopeful designs, too.

The Hippo Roller is a simple device that consists of a cylindrical water tank, turned on its side, with a mounted handle. The canister can easily be rolled across vast distances and terrains, while preserving the purity of the water inside. The design was in response to a humanitarian crisis in South Africa during a period of drought in the late 1990s. Without this basic technology, villagers would have to retrieve water from nearby streams and rivers and carry the water canister by balancing it on top of their heads. The Hippo Roller immediately benefited this population because it afforded them the ability to transport higher volumes of water with less effort. The ingenious mechanism won the South African "Design for Development Award" and the endorsement of the country's former president, Nelson Mandela. Today, there are more than 65,000 Hippo Rollers in use, servicing 56 countries and more than 650,000 people.[6] This is only one of countless examples that demonstrate how design can be used ethically to make a positive difference in the world, and it makes me feel optimistic that the future of design is bright.

On responsibility

Ethical design is a luxury. Many cultures and peoples are not afforded the opportunity to consider whether their infrastructure was informed by race or politics, whether their buildings are eco-minded, or whether their personal technologies and clothing are sustainably designed. Yes, evaluating the concept of ethical design, although it is always present regardless of circumstance, seems opulent when that design is necessary for survival. If you are reading this, then it is likely that you are among those fortunate enough to contemplate ethics in design. We have a choice in how we behave and in the world that we create. As such, we have an even stronger obligation to those less privileged to behave ethically.

In Chapter 2, I introduced a simple definition of *ethics*: mindful well-being. Not everyone shares the same values, virtues, or morals, but we all are governed by ethics. Essentially, ethics is "what ought to be." Ethical behavior serves the common good. By extension, I offered a definition for *ethical design:* a set of standards and practices that guides how an object, process, or system is shaped and how it subsequently affects its users as well as all those in its periphery.

Evaluating things for their ethical design forces us to ask probing questions like: Who benefits from the design? Who suffers from it? What intent lies behind the design? What residual effects might occur from this design? This evaluative process stands in direct opposition to the unofficial mantra of modern tech innovators like those in Silicon Valley, whose rally cry is to "Move fast and break things." Instead, ethical design requires slowness. An ethically designed object is deliberately mindful of its impact on individuals and on society, not just its stakeholders and bottom line. Indeed, many practices and beliefs espoused by indigenous peoples all over the world have long-recognized ethical considerations like sustainability, conservation, and preservation. Their practices – sometimes criticized as unsophisticated – are designed to coexist harmoniously with the natural environment. These cultures work diligently to foster their resources, not exploit them for profit or immediate gratification. Their heritage honors their ancestors and anticipates the world that they wish to leave their descendants.

Ethical design is unselfish. This means that ethical design can be hard. Sometimes, it can mean choosing the common good over our personal desires, sacrificing our own interests for others', and exercising restraint even when it goes unacknowledged. Yet, we have a shared responsibility – a social contract even – to act ethically. Peter-Paul Verbeek said that ethics is "mastering the art of living."[7] Langdon Winner said that we constantly enter into social contracts with everyday objects, and this constitutes the "fabric of our society." Victor Papanek described how everything has a "spiritual value"[8] – or a lack thereof – that can be observed if we look hard enough. And Nicholas Carr reminds us that when we use an object, "Its intentions are our intentions."[9] Ethical design becomes inescapable because it is everywhere and everything, so we must be mindful and observant of how design informs our lives.

Our built environment dictates our behaviors. Plain and simple. This statement is central to the thesis of this book, which is that we are all tethered to the ethical design of our surroundings – whether we are designing a building, consuming products, supporting causes, accepting terms and conditions, or blindly following technological trends. But we are active participants – not just recipients – of these designs, and with participation comes responsibility. I began this book with the example of the K-Cup coffee pod. Admittedly, I was a chronic K-Cup coffee pod user for several years, which means that I singlehandedly contributed thousands of tiny plastic cups to my local landfill. Eventually, I realized that I had a choice. Yes, the coffee pods were more convenient – just pop one into the Keurig machine and get a cup of coffee in a matter of seconds. This seemed preferable to brewing an entire pot of coffee, which took more time and effort. But the more I contemplated ethics and how they are imbued with everything around me, I stopped buying and using K-Cups. I found myself in a sort of silent boycott. And it donned on me; this small act of resistance doesn't need to be small. We all have a voice. We get to decide whether we will purchase fast fashion, consume food that has been factory-farmed, visit roadside zoos, adopt technologies that violate our privacy and exploit our attention, substitute human relationships for the virtual, and endorse lawmakers and politicians who support cruel policies. As I have argued throughout this book, we have an individual responsibility to promote ethics in the designs that we encounter. But this doesn't mean that our actions lead to an immediate change. None of the examples of unethical design in this book can be defeated by one person. But collectively, these voices reach a tipping point at which unethical designs cease to be popular, profitable, and, eventually, justifiable. I also realize it's not that simple. Everyone's morals and values are different, and often, we clash with one another. And that's what makes ethics so special; ethics is what ought to be, not necessarily what we want it to be.

Our objects and public spaces reflect the values, morals, and ethics of the designer and of the cultural and societal contexts under which they were designed. As a result, design becomes a conduit for our own image. Designers have an especially powerful responsibility to the rest of the world, and they are often great influencers, whether intentional or not. But it is not sufficient to simply defer to designers to save the world (or blame them for ruining it). As consumers, we first must be mindful of the macro and micro designs that shape our surroundings, and second, be highly critical of their ethical postures. We must be resolute in making ethical choices but be willing to make calculated compromises when ethical dilemmas arise. So, in the end, we are all contributors to design. Everything that is, and ever was, has been designed. Everything that will be, will be designed. And all this reflects who we are, who we aspire to be, and what we will be remembered for.

Notes

Chapter 1

[1] Winner L. (1986). *The whale and the reactor: a search for limits in an age of high technology.* University of Chicago Press.
[2] Bever, L. (2015, March 5). K-Cups inventor: I feel bad sometimes that I ever did it. *The Washington Post.* https://www.washingtonpost.com/news/morning-mix/wp/2015/03/05/k-cups-inventor-i-feel-bad-sometimes-that-i-ever-did-it/
[3] Wroclawski, D. & Bradley, M. (2024, March 12). *The ultimate guide to recycling K-Cups, Nespresso Capsules, and other coffee pods.* Consumer Reports. https://www.consumerreports.org/home-garden/recycling/recycle-k-cups-nespresso-capsules-coffee-pods-a3276889649/#:~:text=But%20the%20best%20solution%20to,preground%20or%20whole%2Dbean%20coffee
[4] Southern Poverty Law Center. (n.d.). *Westboro Baptist Church.* https://www.splcenter.org/fighting-hate/extremist-files/group/westboro-baptist-church
[5] Baker, J. O., Bader, C. D., & Hirsch, K. (2015). Desecration, Moral Boundaries, and the Movement of Law: The Case of Westboro Baptist Church. *Deviant Behavior, 36*(1), 42–67. https://doi.org/10.1080/01639625.2014.906282
[6] Lee, H. N. (1928). Morals, Morality, and Ethics: Suggested Terminology. *International Journal of Ethics, 38*(4), 450–466. http://www.jstor.org/stable/2377964
[7] Dean, T. (2021, February 2). *Ethics explainer: Moral relativism.* The Ethics Centre. https://ethics.org.au/ethics-explainer-moral-relativism/
[8] Verbeek, P.-P. (2011). *Moralizing technology: understanding and designing the morality of things.* The University of Chicago Press.
[9] Singer, P. (2016). *Ethics in the Real World: 82 Brief Essays on Things That Matter* (REV-Revised). Princeton University Press. https://doi.org/10.2307/j.ctvc77g39
[10] Singer, P. (2006, January 4). *Godless morality.* Project Syndicate. https://www.project-syndicate.org/commentary/godless-morality
[11] Verbeek, P.-P. (2005). *What things do: Philosophical reflections on technology, agency, and design.* Pennsylvania State University Press.
[12] Douglass, B. (1980). The Common Good and the Public Interest. *Political Theory, 8*(1), 103–117. http://www.jstor.org/stable/190769
[13] Reich, R. B. (2018). *The common good.* First edition. New York, Alfred A. Knopf.
[14] Governors Highway Safety Association. (2023, March). *Motorcyclists.* https://www.ghsa.org/state-laws/issues/Motorcyclists#:~:text=Illinois%2C%20Iowa%20and%20New%20Hampshire,not%20have%20motorcycle%20helmet%20laws.
[15] Booz, Mark L. (1986). The constitutionality of mandatory seat belt laws. Virginia Highway & Transportation Research Council. Report # VHTRC 86-R41. Retrieved from https://rosap.ntl.bts.gov/view/dot/19122.

[16] Insurance Institute for Highway Safety. (2023, July). *Seat belt laws.* https://www.iihs.org/topics/seat-belts#:~:text=With%20the%20exception%20of%20New,and%20the%20District%20of%20Columbia.

[17] Holdorf, W.J. (1987, February 5). *Seat belt law an assault on freedom.* The Chicago Tribune. https://www.chicagotribune.com/news/ct-xpm-1987-02-05-8701090791-story.html

[18] Centers for Disease Control and Prevention. (2020, April 28). *Smoking during pregnancy.* https://www.cdc.gov/tobacco/basic_information/health_effects/pregnancy/index.htm

[19] Franklin D. Roosevelt Presidential Library and Museum. (n.d.). *FDR and the Four Freedoms Speech.* https://www.fdrlibrary.org/four-freedoms

[20] Verbeek, P.-P. (2006). Materializing morality: Design ethics and technological mediation. *Science, Technology & Human Values, 31*(3), 361–380. https://doi.org/10.1177/0162243905285847

[21] Sabin Center for Climate Change Law. (2024). *About.* http://climatecasechart.com/about/

[22] Youth v. Gov. (2023). *Historic climate trial: Held v. State of Montana.* https://www.youthvgov.org/held-v-montana

[23] Balkan, A. & Kalbag, L. (n.d.). *Ethical design.* https://ind.ie/ethical-design/

[24] Balkan, A. & Kalbag, L. (n.d.). *Ethical design.* https://ind.ie/ethical-design/

[25] Balkan, A. & Kalbag, L. (n.d.). *Ethical design.* https://ind.ie/ethical-design/

[26] Verbeek, P.-P. (2011). *Moralizing technology: understanding and designing the morality of things.* The University of Chicago Press.

[27] Winner L. (1986). *The whale and the reactor: a search for limits in an age of high technology.* University of Chicago Press.

[28] Friedman, B., & Hendry, D. G. (2019). *Value sensitive design: shaping technology with moral imagination.* MIT Press.

[29] Friedman, B., & Hendry, D. G. (2019). *Value sensitive design: shaping technology with moral imagination.* MIT Press.

[30] Jacobs, N., & Huldtgren, A. (2021). Why value sensitive design needs ethical commitments. *Ethics & Information Technology, 23*(1), 23–26. https://doi.org/10.1007/s10676-018-9467-3

[31] van de Poel, I. (2013). Translating Values into Design Requirements. In Michelfelder, D., McCarthy, N., Goldberg, D. (eds) *Philosophy and Engineering: Reflections on Practice, Principles and Process. Philosophy of Engineering and Technology,* vol 15. Springer, Dordrecht. https://doi-org.proxy1.library.jhu.edu/10.1007/978-94-007-7762-0_20

[32] Coalition for the Homeless. (2023, February). *Basic facts about homelessness: New York City.* https://www.coalitionforthehomeless.org/basic-facts-about-homelessness-new-york-city/

[33] Madden, K. Kent, F., & Peinhardt, K. (2021, November 8). *How seating shapes welcoming cities.* Social Life Project. https://www.sociallifeproject.org/have-a-seat-how-seating-shapes-welcoming-cities/

[34] Madden, K. Kent, F., & Peinhardt, K. (2021, November 8). *How seating shapes welcoming cities.* Social Life Project. https://www.sociallifeproject.org/have-a-seat-how-seating-shapes-welcoming-cities/

[35] Rosenberger, R. (2017). *Callous Objects: Designs Against the Homeless.* Minneapolis, MN: University of Minnesota Press.
[36] Ross, S. (n.d.). *2005-2006 Archisuits.* https://insecurespaces.net/archisuits-2005-2006/
[37] Winner L. (1986). *The whale and the reactor: a search for limits in an age of high technology.* University of Chicago Press.
[38] Mitchell, WJT (1998). The violence of public art. In Ostrow, S. (Ed.), *Beauty is nowhere: ethical issues in art and design.* Routledge. https://doi.org/10.4324/9780203871423
[39] Mitchell, WJT (1998). The violence of public art. In Ostrow, S. (Ed.), *Beauty is nowhere: ethical issues in art and design.* Routledge. https://doi.org/10.4324/9780203871423
[40] Mitchell, WJT (1998). The violence of public art. In Ostrow, S. (Ed.), *Beauty is nowhere: ethical issues in art and design.* Routledge. https://doi.org/10.4324/9780203871423
[41] Kamiya, G. (2021, February 2). *The holier-than-thou crusade in San Francisco.* The Atlantic. https://www.theatlantic.com/ideas/archive/2021/02/san-francisco-renaming-spree/617894/

Chapter 2

[1] Murrow, E. (1955, January 4). *A Conversation with J. Robert Oppenheimer.* See It Now. https://famous-trials.com/oppenheimer/2694-edward-r-murrow-interviews-oppenheimer-see-it-now
[2] The name comes from a notable figure in the design world, Don Norman, whom I will reference throughout this text. "Norman doors" are covered more extensively in Chapter 3.
[3] AtomicArchive.com. (n.d.). *Einstein's letter to President Roosevelt – 1939.* https://www.atomicarchive.com/resources/documents/beginnings/einstein.html
[4] American Museum of Natural History. (n.d.). *The Manhattan Project.* https://www.amnh.org/exhibitions/einstein/peace-and-war/the-manhattan-project
[5] Tully, J. (1986, December 19). *Einstein letter sold to Forbes.* The Washington Post. https://www.washingtonpost.com/archive/lifestyle/1986/12/20/einstein-letter-sold-to-forbes/1f5d23ff-47e3-43ba-a7c0-0199643f1933/
[6] AtomicArchive.com. (n.d.). *Report of the committee on political and social problems.* https://www.atomicarchive.com/resources/documents/manhattan-project/franck-report.html#V
[7] Temperton, J. (2023, July 21). *'Now I am become death, the destroyer of worlds.' The story of Oppenheimer's infamous quote.* Wired. https://www.wired.com/story/manhattan-project-robert-oppenheimer/
[8] J. Robert Oppenheimer, atomic bomb pioneer, dies. (1967, February 19). *The New York Times.* https://archive.nytimes.com/www.nytimes.com/learning/general/onthisday/bday/0422.html
[9] Markkula Center for Applied Ethics. (2021, November 8). *A framework for ethical decision making.* https://www.scu.edu/ethics/ethics-resources/a-framework-for-ethical-decision-making/
[10] IMDB. (n.d.). *Jeff Goldblum: Malcolm.* https://www.imdb.com/title/tt0107290/characters/nm0000156

[11] Tseng, P. E., & Wang, Y. H. (2021). Deontological or utilitarian? An eternal ethical dilemma in *Outbreak. International Journal of Environmental Research and Public Health, 18*(16), 8565. https://doi.org/10.3390/ijerph18168565

[12] Hampton, L. (2023). Jeremy Bentham. In R.Y. Chappell, D. Meissner, and W. MacAskill (eds.), *An introduction to utilitarianism.* https://www.utilitarianism.net/utilitarian-thinker/jeremy-bentham

[13] Driver, J. (2014, September 22). The history of utilitarianism. *The Stanford Encyclopedia of Philosophy* (Winter 2022 Edition). https://plato.stanford.edu/archives/win2022/entries/utilitarianism-history/.

[14] United States Geological Survey. (n.d.). *How many homes can an average wind turbine power?* https://www.usgs.gov/faqs/how-many-homes-can-average-wind-turbine-power

[15] Hersh, M. (2015). Ethical engineering: definitions, theories, and techniques. In M. Hersh (Ed.) *Ethical Engineering for International Development and Environmental Sustainability.* Springer-Verlag.

[16] Soccer Stadium Digest. (n.d.). *2022 MLS attendance.* https://soccerstadiumdigest.com/2022-mls-attendance/

[17] Cardenas, F. (2023, August 4). Why Messi's arrival doesn't mean guaranteed sellouts at Inter Miami's stadium. *The Athletic.* https://theathletic.com/4748102/2023/08/04/inter-miami-ticket-prices-messi/

[18] This online interactive presents variations of the original trolley problem, allowing you to decide and then see the percentage of others who agreed / disagreed with your decision: https://neal.fun/absurd-trolley-problems/

[19] Bakewell, S. (2013, November 22). *Clang went the trolley.* The New York Times. https://www.nytimes.com/2013/11/24/books/review/would-you-kill-the-fat-man-and-the-trolley-problem.html

[20] Lillehammer, H. (2023). Keeping track of your trolleys: origins and destinations. In H. Lillehammer (Ed.), *The trolley problem* (Classic Philosophical Arguments, pp. 7-24). Cambridge: Cambridge University Press.

[21] Gallop, J.D. (2018, June 22). *No charges for 5 teens who mocked and filmed drowning man, Jamel Dunn, in a Cocoa pond.* Florida Today. https://www.floridatoday.com/story/news/2018/06/22/no-charges-year-after-teens-mocked-drowning-man-prosecutors-rule-out-filing-charges/723259002/

[22] Unless, of course, there is a "special relationship" where one person is dependent on the other, such as in the case of a parent-child or a teacher-student relationship. Then, this burden is greater for the care provider.

[23] Cowell, A. (2008, April 8). *Negligent driving killed Diana, jury finds.* The New York Times. https://www.nytimes.com/2008/04/08/world/europe/08diana.html

[24] Lloyd, S. & Sreedhar, S. (2022, September 12). Hobbes's moral and political philosophy. *The Stanford Encyclopedia of Philosophy* (Winter 2022 Edition). https://plato.stanford.edu/entries/hobbes-moral/#StaNat

[25] Papanek, V. (1998). Toward the spiritual in design. In Ostrow, S. (Ed.), *Beauty is nowhere: ethical issues in art and design.* Routledge. https://doi.org/10.4324/9780203871423

[26] Gogoll, J. et al. (2021). Ethics in the software development process: from codes of conduct to ethical deliberation. *Philos. Technol, 34,* 1085–1108.

[27] They won the World Series in 2016, not 2015 as Zemeckis – a Chicagoan – facetiously predicted in the film.

[28] Dunne, A. & Raby, F. (n.d.). *Critical design FAQ.* https://dunneandraby.co.uk/content/bydandr/13/0

[29] Critical design FAQ. (n.d.). *Reading Design.* https://www.readingdesign.org/critical-design-faq

[30] Dunne, A. (1999). *Hertzian tales: electronic products, aesthetic experience, and critical design.* MIT Press.

[31] Speculative EDU. (2019, July 23). *Critical about critical and speculative design.* https://speculativeedu.eu/critical-about-critical-and-speculative-design/

[32] Chakrabarty, P. (2023, January 24). The basics you need to know & understand about speculative design. *Medium.* https://uxplanet.org/the-basics-you-need-to-know-and-understand-about-speculative-design-8a8bf5be4162

[33] Overkamp, L. (2019, May 30). *Daily ethical design.* A List Apart. https://alistapart.com/article/daily-ethical-design/

[34] Polakova, S. & Cruchade, A. (2018, October 24). *Can speculative design make UX better?* Medium. https://uxplanet.org/can-speculative-design-make-ux-better-design-trend-4-4-ce8d13148e5d

[35] Interaction 18. (n.d.). *The Oppenheimer Moment.* https://interaction18.ixda.org/program/keynote--alan-cooper/

[36] Pratten, J.D., Wright, D.L. (2013). Unethical products. In: Idowu, S.O., Capaldi, N., Zu, L., Gupta, A.D. (eds) *Encyclopedia of Corporate Social Responsibility.* Springer, Berlin, Heidelberg. https://doi.org/10.1007/978-3-642-28036-8_598

[37] Harari, Y. N. (2024). *Nexus: a brief history of information networks from the Stone Age to AI.* First U.S. edition. New York, Random House.

[38] Perez, S. (2021, September 16). *Tile secures $40 million to take on Apple AirTag with new products.* Tech Crunch. https://techcrunch.com/2021/09/16/tile-secures-another-40-million-to-take-on-apple-airtag-with-new-products/

[39] Axon, S. (2021, November 23). Bluetooth tracking company Tile acquired for $205 million. *Ars Technica.* https://arstechnica.com/gadgets/2021/11/bluetooth-tracking-company-tile-acquired-for-205-million/

[40] Viacorka, F. [@franaviacorka] (2022, April 5). *Ukrainians are locating their devices on the territory of the Homiel region, Belarus, where part of the Russian army retreated.* [Post]. X. https://twitter.com/franakviacorka/status/1511297092836278283; Fowler, G. (2022, March 31). *Am I being tracked? Anti-stalking tech from Apple, Tile falls short.* The Washington Post. https://www.washingtonpost.com/technology/2022/03/31/airtags-stalking/

[41] Fowler, G. (2022, March 31). *Am I being tracked? Anti-stalking tech from Apple, Tile falls short.* The Washington Post. https://www.washingtonpost.com/technology/2022/03/31/airtags-stalking/

[42] Moore, A. (2022, September 5). *'I didn't want it anywhere near me': how the Apple AirTag became a gift to stalkers.* The Guardian. https://www.theguardian.com/technology/2022/sep/05/i-didnt-want-it-anywhere-near-me-how-the-apple-airtag-became-a-gift-to-stalkers

[43] Morgan, R.E. & Truman, J.L. (2022, February). *Stalking victimization, 2019.* U.S. Department of Justice. https://bjs.ojp.gov/content/pub/pdf/sv19.pdf

[44] Apple introduces AirTag. (2021, April 20). *Apple.* https://www.apple.com/newsroom/2021/04/apple-introduces-airtag/

[45] Mason, R. O. (1986). Four Ethical Issues of the Information Age. *MIS Quarterly, 10*(1), 5–12. https://doi.org/10.2307/248873

[46] Merchant, N. (n.d.). *IoT technologies explained: History, examples, risks, & future.* Vision of Humanity. https://www.visionofhumanity.org/what-is-the-internet-of-things/#:~:text=The%20term%20'Internet%20of%20Things,them%20through%20a%20supply%20chain.

[47] Mason, R. O. (1986). Four Ethical Issues of the Information Age. *MIS Quarterly, 10*(1), 5–12. https://doi.org/10.2307/248873

[48] Mason, R. O. (1986). Four Ethical Issues of the Information Age. *MIS Quarterly, 10*(1), 5–12. https://doi.org/10.2307/248873

Chapter 3

[1] Churchill, W. (1943, October). *House of Commons Rebuilding.* https://api.parliament.uk/historic-hansard/commons/1943/oct/28/house-of-commons-rebuilding

[2] Winner, L. (2014). Technologies as forms of life. In: Sandler, R.L. (eds) *Ethics and emerging technologies.* Palgrave Macmillan, London.

[3] Dotov, D. G., Nie, L., & Chemero, A. (2010). A demonstration of the transition from ready-to-hand to unready-to-hand. *PLoS ONE, 5*(3), 1–9.

[4] Aagaard, J. (2021). The passive subject: a phenomenological contribution to STS. *Social Epistemology Review and Reply Collective 10*(5): 14-19. https://wp.me/p1Bfg0-5Pb

[5] Nelson, G. (1977). *How to see: visual adventures in a world God never made* (1st ed). Little Brown.

[6] Stein, S.M. & Harper, T.L. (2013). Designing for meaning: the ethical designer's responsibility. In C. Basta & S. Moroni (Eds.), *Ethics, design and planning of the built environment* (pps. 143-166). Springer Dordrecht.

[7] Lockton, D. (n.d.). *Architectures of control in the built environment.* Architectures. https://architectures.danlockton.co.uk/architectures-of-control-in-the-built-environment/

[8] Rybczynski, W. (2013). *How architecture works: A humanist's toolkit.* Farrar, Strous, and Giroux. New York.

[9] Alexander, C. (1979). *The timeless way of building.* Oxford University Press, New York.

[10] Anthes, E. (2020). *The great indoors: the surprising science of how buildings shape our behavior, health, and happiness* (First Picador paperback edition). Scientific American/Picador.

[11] Gibson, J.J. (1979). *The ecological approach to visual perception.* Boston: Houghton Mifflin.

[12] Gibson, J.J. (1979). *The ecological approach to visual perception.* Boston: Houghton Mifflin.
[13] Gibson, J.J. (1979). *The ecological approach to visual perception.* Boston: Houghton Mifflin.
[14] Forcing functions. (n.d.). *Interaction Design Foundation.* https://www.interaction-design.org/literature/book/the-glossary-of-human-computer-interaction/forcing-functions
[15] Norman, D. A. (1988). *The psychology of everyday things.* Basic Books.
[16] Salvador Dali: The persistence of memory. (n.d.). *Museum of Modern Art.* https://www.moma.org/collection/works/79018
[17] René Magritte: Belgian, 1898-1967. (n.d.). *Museum of Modern Art.* https://www.moma.org/artists/3692#fn:1
[18] A portfolio of Kamprani's work is featured on her site, theuncomfortable.com.
[19] Kamprani, K. (n.d.). *The Uncomfortable.* https://www.theuncomfortable.com/about/
[20] Ong, P. (n.d.). *Priscilla Ong studio.* https://www.priscillaong.com/
[21] de Boer. (2021). Explaining multistability: postphenomenology and affordances of technologies. *AI & Society.* https://doi.org/10.1007/s00146-021-01272-3
[22] Lo, A. (2017, December 21). *The debate: Is hostile architecture designing people – and nature – out of cities?* CNN. https://www.cnn.com/style/article/new-dean-harvey-james-furzer-hostile-architecture-debate/index.html
[23] Watch the video demonstration of Brunsing's piece here: https://vimeo.com/1665301
[24] Rosenberger, R. (2017). *Callous objects: designs against the homeless.* Minneapolis, MN: University of Minnesota Press.
[25] Olmos, S. (2023, April 23). *LA is blasting classical music to deter homeless people from gathering at metro stations.* NPR. https://www.npr.org/2023/04/23/1171494452/la-is-blasting-classical-music-to-deter-homeless-people-from-gathering-at-metro-
[26] How to stop anti-social behavior. (n.d.). *Mosquito loitering solutions.* https://mosquitoloiteringsolutions.com/why-mosquito/anti-social-behaviour/
[27] How does the Mosquito alarm work? (n.d.). *Mosquito loitering solutions.* https://mosquitoloiteringsolutions.com/faq/how-does-it-work/
[28] How Mosquito helps in the community. (n.d.). *Mosquito loitering solutions.* https://mosquitoloiteringsolutions.com/how-mosquito-has-helped/in-the-community/
[29] Gregersen, L. (2022, August 11). *How a small Canadian city took on chronic homelessness.* Next City. https://nextcity.org/urbanist-news/medicine-hat-alberta-canada-city-chronic-homelessness
[30] Fraieli, A. (2021, May 10). *The cost to criminalize homelessness.* The Homeless Voice. https://homelessvoice.org/the-cost-to-criminalize-homelessness/
[31] Caro, R.A. (1974). *The power broker: Robert Moses and the fall of New York.* Knopf: New York.
[32] The New York Historical Society displays an exhibit devoted to Robert Moses and his racist parkway. This includes the tally sheet that Caro and his wife, Ina, kept.
[33] Lung, N. (2023, October 20). *Most New York Airbnb applications don't meet city requirements.* Bloomberg. https://www.bloomberg.com/news/articles/2023-10-20/most-new-york-airbnb-applications-don-t-meet-city-requirements

[34] Kessler, G. (2021, November 10). *Robert Moses and the sage of the racist parkway bridges. The* Washington Post. https://www.washingtonpost.com/politics/2021/11/10/robert-moses-saga-racist-parkway-bridges/

[35] Kessler, G. (2021, November 10). *Robert Moses and the sage of the racist parkway bridges.* The Washington Post. https://www.washingtonpost.com/politics/2021/11/10/robert-moses-saga-racist-parkway-bridges/

[36] Baba, A. (2021, November 9). *Buttigieg is right – America's roads are racist. Ted Cruz and Tucker Carlson are pretending to misunderstand.* Independent. https://www.independent.co.uk/voices/buttigieg-racist-ted-cruz-tucker-carlson-b1954576.html

[37] Dittmer, J. (1997). "Too busy to hate": Race, class, and politics in twentieth-century Atlanta. *The Georgia Historical Quarterly, 81*(1), 103–117.

[38] *The Atlanta Constitution.* https://dilemmaxdotnet.files.wordpress.com/2021/02/1962-december-19-atlanta-wall-03.jpg

[39] Collins, W. J., & Shester, K. L. (2013). Slum clearance and urban renewal in the United States. *American Economic Journal: Applied Economics, 5*(1), 239–273. http://www.jstor.org/stable/43189425

[40] Pandya, D. & Kalra, A. (2023, July 20). *India's Adani to convert Mumbai slum into modern city despite challenges.* Reuters. https://www.reuters.com/world/india/indias-adani-convert-mumbai-slum-into-modern-city-despite-challenges-2023-07-20/

[41] Slum clearance. (1933, December 19). *The New York Times.* https://www.nytimes.com/1933/12/19/archives/slum-clearance.html?url=http%3A%2F%2Ftimesmachine.content-tagging.us-east-1-01.prd.dvsp.nyt.net%2Ftimesmachine%2F1933%2F12%2F19%2F105832327.html

[42] Parshina-Kottas, Y., Singhvi, A., Burcj, A., Griggs, T., Grondahl, M., Huang, L., Wallace, T., White, J., & Williams, J. (2021, May 24). *What the Tulsa race massacre destroyed.* The New York Times. https://www.nytimes.com/interactive/2021/05/24/us/tulsa-race-massacre.html

[43] Moreno, C. (2021, June 21). *Decades after the Tulsa race massacre, urban 'renewal' sparked Black Wall Street's second destruction.* Smithsonian Magazine. https://www.smithsonianmag.com/history/black-wall-streets-second-destruction-180977871/

[44] Sullivan, A. (2021, May 25). *U.S. freeways flattened Black neighborhoods nationwide.* Reuters. https://www.reuters.com/world/us/us-freeways-flattened-black-neighborhoods-nationwide-2021-05-25/

[45] Structural racism and discrimination. (2023, February 23). *National Institute of Minority Health and Health Disparities.* https://www.nimhd.nih.gov/resources/understanding-health-disparities/srd.html

[46] Thoreau, H. D. (1917). *Walden, or, Life in the woods.* Chicago, New York: Scott, Foresman and Company.

[47] Hickman, M. (2020, November 2). *Copenhill, BIG's skiable waste-to-energy power plant, gets sweeping new photos from Hufton + Crow.* The Architect's Newspaper. https://www.archpaper.com/2020/11/copenhill-bigs-skiable-waste-to-energy-power-plant-gets-sweeping-new-photos-from-hufton-crow/

[48] Fisher, T. (2019). *The architecture of ethics.* Routledge.

[49] Helliwell, J. F., Huang, H., Norton, M., Goff, L., & Wang, S. (2023). World happiness, trust, and social connections in times of crisis. In *World Happiness Report 2023* (11th ed., Chapter 2). Sustainable Development Solutions Network. https://worldhappiness.report/ed/2023/world-happiness-trust-and-social-connections-in-times-of-crisis/#ranking-of-happiness-2020-2022

[50] What is NEOM? (n.d.). *NEOM*. https://www.neom.com/en-us/about

[51] Boeck, A. (2023, June 27). Why Saudi Arabia's "The Line" isn't a revolution in urban living. *EurekAlert!* https://www.eurekalert.org/news-releases/993409

[52] Akkad, D. (2022, October 7). Neom: Saudi Arabia sentences tribesmen to death for resisting displacement. *Middle East Eye*. https://www.middleeasteye.net/news/neom-saudi-arabia-sentences-tribesmen-death-resisting-displacement

Chapter 4

[1] Carr, N. (2011). *The shallows*. WW Norton.

[2] Dunham, W. (2022, July 19). Scientists uncover history of 'ridiculously charming' penguins. *Reuters*. https://www.reuters.com/lifestyle/science/scientists-uncover-history-ridiculously-charming-penguins-2022-07-19/#:~:text=The%20researchers%20said%20penguins%20evolved,becoming%20excellent%20swimmers%20and%20divers.

[3] Carnegie Science Earth and Planets Laboratory. (2023, October 16). Scientists, philosophers identify nature's missing evolutionary law. *ScienceDaily*. https://www.sciencedaily.com/releases/2023/10/231016163100.htm

[4] Kampourakis, K. (2020). Students' "teleological misconceptions" in evolution education: why the underlying design stance, not teleology per se, is the problem. *Evo Edu Outreach, 13*(1). https://doi.org/10.1186/s12052-019-0116-z

[5] Jones, R.P. & Cox, D. (2011). *Few Americans see natural disasters a sign from god*. PRRI. http://www.prri.org/research/few-americans-see-earthquakes-floods-and-other-natural-disasters-a-sign-from-god-2/.

[6] Shimron, Y. (2023, October 6). *U.S. faith groups do not view climate change as a crisis, new poll finds*. The Washington Post. https://www.washingtonpost.com/religion/2023/10/06/religion-climate-change-survey/

[7] Bejan, A., & Lorente, S. (2010). The constructal law of design and evolution in nature. *Philosophical transactions of the Royal Society of London. Series B, Biological sciences, 365*(1545), 1335–1347. https://doi.org/10.1098/rstb.2009.0302

[8] Bejan, A. & Zane, J.P. (2012). *Design in nature*. Knopf Doubleday Publishing Group.

[9] United Nations. (2021, October 3). *UN chief promotes 'enormous' benefits of greener cities*. https://news.un.org/en/story/2021/10/1101992

[10] Architecture 2030. (n.d.). *Why the built environment?* https://www.architecture2030.org/why-the-built-environment/

[11] Architecture 2030. (n.d.). *Why the built environment?* https://www.architecture2030.org/why-the-built-environment/

[12] AIA. (2023, November 29). *Climate action plan*. https://www.aia.org/resources/6307290-climate-action-plan

[13] Murray, C. (2019, March 28). *It's time for architects to choose ethics over aesthetics.* dezeen. https://www.dezeen.com/2019/03/28/opinion-christine-murray-climate-change/

[14] Koonin, S. (2021). *Unsettled: What climate science tells us, what it doesn't, and why it matters.* BenBella Books, Inc.: Dallas.

[15] Koonin, S. (2021). *Unsettled: What climate science tells us, what it doesn't, and why it matters.* BenBella Books, Inc.: Dallas.

[16] The Associated Press. (2022, August 21). *China plans to use chemicals to generate rain to protect its grain harvest.* NPR. https://www.npr.org/2022/08/21/1118683699/china-grain-harvest-cloud-seeding

[17] Pontin, M.W. (2008, March 25). *Weather engineering in China.* MIT Technology Review. https://www.technologyreview.com/2008/03/25/270084/weather-engineering-in-china/

[18] Harvey, C. (2021, March 16). *Eight states are seeding clouds to overcome megadrought.* Scientific American. https://www.scientificamerican.com/article/eight-states-are-seeding-clouds-to-overcome-megadrought/

[19] Gray, R. (2007, April 22). *How we made the Chernobyl rain.* The Telegraph. https://www.telegraph.co.uk/news/worldnews/1549366/How-we-made-the-Chernobyl-rain.html

[20] EPA. (n.d.). *Climate change and social vulnerability in the United States: a focus on six impacts.* U.S. Environmental Protection Agency. https://www.epa.gov/cira/social-vulnerability-report

[21] Sagan, C. (1985). *We are all in this greenhouse together.* [Video]. YouTube. https://www.youtube.com/watch?v=EuR9CzWcCfc

[22] According to an analysis report published by *The Carbon Brief*, the countries with the largest cumulative emissions from 1850-2021 are United States, China, and Russia, respectively.

[23] United Nations. (n.d.). *Sustainability.* https://www.un.org/en/academic-impact/sustainability

[24] Barr, S. (2007). Factors influencing environmental attitudes and behaviors: a U.K. case study of household waste management. *Environment and Behavior, 39*(4), 435-473. https://doi.org/10.1177/0013916505283421

[25] Project for Public Spaces. (2008, December 31). *Waste receptacles.* https://www.pps.org/article/wastereceptacles

[26] Recycling Today. (2019, November). *The woes of poorly sorted waste – an industry perspective.* https://www.recyclingtoday.com/article/the-woes-of-poorly-sorted-waste--an-industry-perspective/

[27] Cooley, A. (2016, July 1). *Design is key to making recycling work.* National Recreation and Park Association. https://www.nrpa.org/parks-recreation-magazine/2016/july/design-is-key-to-making-recycling-work/

[28] Wang, Y. & Shi, Q. (2022). The impact of municipal solid waste sorting policy on air pollution: Evidence from Shanghai, China. *PLoS ONE, 17*(11). https://doi.org/10.1371/journal.pone.0277035

[29] Zink, T., & Geyer, R. (2016). There is no such thing as a green product. *Stanford Social Innovation Review, 14*(2), 26–31. https://doi.org/10.48558/G401-WQ16

[30] Dahmus, J.B. (2014). Can efficiency improvements reduce resource consumption? A historical analysis of ten activities. *Journal of Industrial Ecology, 18*(6), 883–897.

[31] European Parliament. (2023, April 27). *Ending fast fashion: tougher rules to fight excessive production and consumption.* https://www.europarl.europa.eu/news/en/press-room/20230424IPR82040/ending-fast-fashion-tougher-rules-to-fight-excessive-production-and-consumption

[32] The Ngram Viewer is an imperfect analysis tool. It only searches a corpus of text that has been digitized into Google Books. So, it is entirely possible (and likely) that the phrase appears earlier in social and cultural writings but not in academic writings.

[33] Similarweb. (n.d.). *Top websites ranking.* https://www.similarweb.com/top-websites/lifestyle/fashion-and-apparel/

[34] Testa, J. (2022, September 1). *The people's republic of Shein.* The New York Times. https://www.nytimes.com/2022/09/01/style/shein-clothing.html

[35] UN Environment Programme. (2022, September 30). *Why you should rethink your next fashion purchase.* https://www.unep.org/news-and-stories/story/why-you-should-rethink-your-next-fashion-purchase

[36] Including myself. I admit that I have made numerous online purchases at SHEIN.com. Admittedly, it's easier to write about ethics than to practice ethical behavior, but I am making a conscious effort to be a better consumer moving forward.

[37] Kennedy, J. (1962, September). *We choose to go to the moon.* Rice University. https://www.rice.edu/jfk-speech

[38] O'Brien, M. (2022, September 2). *Years late and billions over budget, NASA's most powerful rocket finally set for takeoff.* PBS. https://www.pbs.org/newshour/show/years-late-and-billions-over-budget-nasas-most-powerful-rocket-finally-set-for-takeoff

[39] Artemis. (n.d.). *The Artemis missions.* NASA. https://www.nasa.gov/specials/artemis/#why

[40] Rose, C. (2015, March 22). *Astrophysicist Neil deGrasse Tyson's one-man mission.* CBS News. https://www.cbsnews.com/news/neil-degrasse-tyson-astrophysicist-charlie-rose-60-minutes/

[41] deGrasse Tyson, N. (2012). The case for space: why we should keep reaching for the stars. *Foreign Affairs, 91*(2), 22–33. http://www.jstor.org/stable/23217218

[42] Hall, L. (2019, July 15). *Going to the moon was hard – but the benefits were huge, for all of us.* NASA. https://www.nasa.gov/technology/tech-transfer-spinoffs/going-to-the-moon-was-hard-but-the-benefits-were-huge-for-all-of-us/

[43] deGrasse Tyson, N. (2023). *What taxpayers gained going to the moon.* [Video]. YouTube. https://www.youtube.com/watch?v=Oe1NyLWGmVI

[44] United States Environmental Protection Agency. (2024, April 10). *EPA history.* https://www.epa.gov/history

[45] McDowell, J. (2024, April 14). *GCAT: general catalog of artificial space objects.* https://planet4589.org/space/gcat/index.html

[46] U.S. Department of Transportation. (2023, September 23). *Federal Aviation Administration.* https://www.faa.gov/sites/faa.gov/files/Report_to_Congress_Reentry_Disposal_of_Satellites.pdf

[47] Murphy, D. et al. (2023). Metals from spacecraft reentry in stratospheric aerosol particles. *PNAS, 120*(43). https://www.pnas.org/doi/full/10.1073/pnas.2313374120
[48] Jordan, D. (2023, September 8). *Elon Musk says he withheld Starlink over Crimea to avoid escalation.* BBC News. https://www.bbc.com/news/world-europe-66752264
[49] ScienceDaily. (2023, April 7). *Internet access must become human right or we risk ever-widening inequality.* https://www.sciencedaily.com/releases/2023/04/230407094243.htm
[50] Central Park Conservancy. (n.d.). *About us.* https://www.centralparknyc.org/about
[51] Foust, J. (2019, April 16). Pepsi drops plans to use orbital billboard. *Space News.* https://spacenews.com/pepsi-drops-plans-to-use-orbital-billboard/
[52] Dickerson, B. (2019, May 30). The history of 20th-century (and earlier) wave pools. *WavePoolMag.* https://wavepoolmag.com/the-complete-history-of-20th-century-wave-pools/#:~:text=In%20the%201870s%20Ludwig%20II,escape%20the%20pressures%20of%20kinghood.
[53] For this section, the term 'artificial wave pool' refers to large-scale artificial wave pools, such as the Surf Ranch and others that utilize large bodies of water, rather than compacted designs, such the FlowRider Surf Simulator, which use jet propulsion to force water out and under a stationary surfer balancing on a board.
[54] Dickerson, B. (2021, April 9). Deep dive: so how much energy does a wave pool use? *WavePoolMag.* https://wavepoolmag.com/deep-dive-so-how-much-energy-does-a-wave-pool-use/#:~:text=300%20to%20400%20kWh%20per,high%20in%20the%20same%20timeframe.
[55] Dickerson, B. (2021, April 9). Deep dive: so how much energy does a wave pool use? *WavePoolMag.* https://wavepoolmag.com/deep-dive-so-how-much-energy-does-a-wave-pool-use/#:~:text=300%20to%20400%20kWh%20per,high%20in%20the%20same%20timeframe; The University of Texas at Austin Energy Institute. *The timeline and events of the February 2021 Texas electric grid blackouts.* https://energy.utexas.edu/research/ercot-blackout-2021.
[56] Surfrider Foundation. (n.d.). *Wave pools: environmental concerns outweigh their value.* https://www.surfrider.eu/learn/news/wave-pools-environmental-concerns-outweigh-their-value/
[57] Plump, W. (n.d.). *Peter Singer on ethical politics.* Princeton Magazine. https://www.princetonmagazine.com/peter-singer-on-ethical-politics/#:~:text=As%20both%20a%20surfer%20and,waves%20are%20good%20or%20not.

Chapter 5

[1] Anthes, E. (2020). *The great indoors: the surprising science of how buildings shape our behavior, health, and happiness* (First Picador paperback edition). Scientific American/Picador.
[2] University of Michigan. (2019, June 20). *Moral concerns override desire to profit from finding a lost wallet.* ScienceDaily. www.sciencedaily.com/releases/2019/06/190620153447.htm

[3] University of Michigan. (2019, June 20). *Moral concerns override desire to profit from finding a lost wallet.* ScienceDaily. www.sciencedaily.com/releases/2019/06/190620153447.htm

[4] Ariely, D. (2013). *The (Honest) Truth About Dishonesty.* HarperCollins.

[5] Bigelow, K. (1991). *Point Break.* Twentieth Century Fox.

[6] Manaugh, G. (2016). *A burglar's guide to the city* (First edition). Farrar, Straus and Giroux.

[7] LAPD. (2023, March 14). *Officer involved shooting occurs during bait car operation.* https://www.lapdonline.org/newsroom/officer-involved-shooting-occurs-during-bait-car-operation/

[8] ABC News. (2008, June 5). *Car bait program goes awry as thief steals car, kills woman.* ABC News. https://abcnews.go.com/GMA/story?id=5004251&page=1

[9] Klosterman, C. (2014, June 20). *So you think you can steal?* The New York Times Magazine. https://www.nytimes.com/2014/06/22/magazine/so-you-think-you-can-steal.html

[10] In 1971, Abbie Hoffman wrote a book that was described as "a manual for survival in the counterculture." It details how to score free food, lodging, transportation, even drugs, and gives specific, actionable advice on how to be a social anarchist. The book was aptly titled *Steal This Book*, and more than 30 publishing houses rejected it because of its cover design. Publishers argued that the title of the book advocated for and encouraged shoplifting (which, aside from being unethical, also is not a very strong revenue model for publishers).

[11] Manaugh, G. (2016). *A burglar's guide to the city* (First edition). Farrar, Straus and Giroux.

[12] The International Crime Prevention Through Environmental Design Association. (n.d.). *What is CPTED?* https://www.cpted.net/

[13] Goldberg, N. (2023, May 8). *Column: high-speed, high adrenaline car chases are an L.A. staple. But are they worth the risks?* Los Angeles Times. https://www.latimes.com/opinion/story/2023-05-08/los-angeles-police-department-car-chase-police-pursuit-deaths-injuries

[14] Bevis, C. & J. Nutter (1978). *Changing street layouts to reduce residential burglary.* Minneapolis: Minnesota Crime Prevention Center.

[15] Bevis, C. & J. Nutter (1978). *Changing street layouts to reduce residential burglary.* Minneapolis: Minnesota Crime Prevention Center.

[16] Webster, C. & Glasze, G. (2002). The global spread of gated communities. *Environment and Planning B: Planning and Design, 29*(3). 315-320. https://www.researchgate.net/publication/23541307_The_Global_Spread_of_Gated_Communities

[17] Larrea, M. (2023, February 13). *What is a gated community? Pros & cons of living in gated communities.* Palm Paradise Realty Group. https://www.palmparadiserealestate.com/blog/gated-community-pros-cons/

[18] Blakely, E. J. & Snyder, M. G. (n.d.). *Fortress America: gated communities in the United States. The New York Times.* https://archive.nytimes.com/www.nytimes.com/books/first/b/blakely-fortress.html

[19] Benjamin, R. (2012, March 29). *The gated community mentality.* The New York Times. https://www.nytimes.com/2012/03/30/opinion/the-gated-community-mentality.html

[20] Chelsea Clip. (n.d.). *How it works.* https://www.chelseaclip.co.uk/pages/how-it-works
[21] Dick, P. K. (1956). The Minority Report. *Fantastic Universe Science Fiction.*
[22] Liberty Human Rights. (2019, February 4). *Liberty report exposes police forces' use of discriminatory data to predict crime.* https://www.libertyhumanrights.org.uk/issue/liberty-report-exposes-police-forces-use-of-discriminatory-data-to-predict-crime/
[23] Fredrickson, D. & Siljander, R. (2002). *Racial profiling.* U.S. Department of Justice Office of Justice Programs. https://www.ojp.gov/ncjrs/virtual-library/abstracts/racial-profiling-0
[24] ACLU. (2005, November 23). *Racial profiling: definition.* https://www.aclu.org/documents/racial-profiling-definition
[25] Gaebler, J., Cai, W., & Goel, S. (2022, September 15). *Police stop Black drivers more often than Whites. We found out why.* The Washington Post. https://www.washingtonpost.com/politics/2022/09/15/driving-while-black-racial-discrimination-traffic-tickets/
[26] NYCLU. Stop-and-frisk data. *ACLU of New York.* https://www.nyclu.org/en/stop-and-frisk-data
[27] Speri, A. (2021, June 10). *The NYPD is still stopping and frisking black people at disproportionate rates.* The Intercept. https://theintercept.com/2021/06/10/stop-and-frisk-new-york-police-racial-disparity/
[28] Heaven, W. (2020, July 17). *Predictive policing algorithms are racist. They need to be dismantled.* MIT Technology Review. https://www.technologyreview.com/2020/07/17/1005396/predictive-policing-algorithms-racist-dismantled-machine-learning-bias-criminal-justice/
[29] Beck, B., Holder, E., Novak, A. & Kaplan, J. (2023). The material of policing: budgets, personnel, and the United States' misdemeanor arrest decline. *The British Journal of Criminology, 63*(2). 330-347. https://academic.oup.com/bjc/article-abstract/63/2/330/6568129?redirectedFrom=fulltext&login=false
[30] Nikke Asia. (2023, July 7). *Japan police to test AI-equipped cameras in protecting VIPs.* https://asia.nikkei.com/Politics/Japan-police-to-test-AI-equipped-cameras-in-protecting-VIPs
[31] Burton-Harris, V. & Mayor, P. (2020, June 24). Wrongfully arrested because face recognition can't tell black people apart. *ACLU.* https://www.aclu.org/news/privacy-technology/wrongfully-arrested-because-face-recognition-cant-tell-black-people-apart
[32] ACLU. (2023, August 6). *After third wrongful arrest, ACLU slams Detroit police department for continuing to use faulty facial recognition technology.* https://www.aclu.org/press-releases/after-third-wrongful-arrest-aclu-slams-detroit-police-department-for-continuing-to-use-faulty-facial-recognition-technology
[33] Fair Trials. (2023, June 16). *EU Parliament approves landmark AI law.* https://www.fairtrials.org/articles/news/eu-parliament-approves-landmark-ai-law/
[34] ACLUSoCal. (2022, October). *Reimagining community safety in California.* https://catalyst-ca.cdn.prismic.io/catalyst-ca/126c30a8-852c-416a-b8a7-55a90c77a04e_APCA+ACLU+REIMAGINING+COMMUNITY+SAFETY+2022_5.pdf
[35] Kang-Brown, J. & Subramanian, R. (2017, June). *Out of sight: the growth of jails in rural America.* Vera Institute of Justice. https://www.vera.org/downloads/publications/out-of-sight-growth-of-jails-rural-america.pdf

[36] Huling, T. (2002). *Invisible punishment: the collateral consequences of mass imprisonment.* The New Press. https://www.prisonpolicy.org/scans/building.html

[37] U.S. General Services Administration. *Alton Lennon federal building and U.S. courthouse, Wilmington, NC.* https://www.gsa.gov/real-estate/historic-preservation/explore-historic-buildings/find-a-building/all-historic-buildings/alton-lennon-federal-building-and-us-courthouse#:~:text=At%20the%20turn%20of%20the,entering%20or%20leaving%20the%20port.

[38] Barlow, A. (n.d.). *Architecture styles in North Carolina courthouses.* North Carolina Administrative Office of the Courts. https://www.nccourts.gov/assets/documents/publications/Brief-History-Architecture-NC-Courthouses.pdf?XMxou7Lols3FqbC5XrzqG.LpWZocCQZR

[39] Lidge, E.F. (2006). Perp walks and prosecutorial ethics. *Nevada Law Review, 7*(26). https://scholars.law.unlv.edu/cgi/viewcontent.cgi?article=1419&context=nlj

[40] Haberman, C. (2018, December 2). *For shame: a brief history of the perp walk.* The New York Times. https://www.nytimes.com/2018/12/02/us/perp-walk.html

[41] HOK. (n.d.). *The future of courthouse design.* https://www.hok.com/ideas/research/the-future-of-courthouse-design/

[42] Rodgers, B. (2023). *Cloak of charity: studies in eighteenth-century philanthropy.* Taylor & Francis Group.

[43] Howard, J. (1777). *The state of the prisons in England and Wales: with preliminary observations, and an account of some foreign prisons.* Printed by William Eyres, MDCCLXXVII.

[44] Forsythe, B. (1984). Prisons and panopticons. *Social Policy & Administration, 18*(1), 68-88.

[45] Bentham, J. (1791). *Panopticon or the Inspection House.* Payne: London.

[46] Foucault M. (1977). *Discipline and punish: the birth of the prison* (1st American). Pantheon Books.

[47] Engstrom, K. V., & van Ginneken, E.F.J.C. (2022). Ethical prison architecture: a systematic literature review of prison design features related to wellbeing. *Space and Culture, 25*(3), 479-503. https://doi.org/10.1177/12063312221104211

[48] Murphy, T. (2015, July 22). *Ben Carson says prison is so comfy some people never want to leave.* Mother Jones. https://www.motherjones.com/politics/2015/07/ben-carson-says-prisons-are-too-comfy/

[49] Chen, M. K., & Shapiro, J. M. (2007). Do harsher prison conditions reduce recidivism? A discontinuity-based approach. *American Law and Economics Review, 9*(1), 1–29. http://www.jstor.org/stable/42705508

[50] Tobón, S. (2022). Do better prisons reduce recidivism? Evidence from a prison construction program. *The Review of Economics and Statistics,104*(6), 1256–1272. doi: https://doi-org.proxy1.library.jhu.edu/10.1162/rest_a_01007

[51] Benko, J. (2015, March 26). The radical humaneness of Norway's Halden Prison. *The New York Times.* https://www.nytimes.com/2015/03/29/magazine/the-radical-humaneness-of-norways-halden-prison.html

[52] United Nations. (2011, October 18). *Solitary confinement should be banned in most cases, UN expert says.* https://news.un.org/en/story/2011/10/392012

[53] Fenster, A. (2020, October 13). New data: solitary confinement increases risk of premature death after release. *Prison Policy Initiative.* https://www.prisonpolicy.org/blog/2020/10/13/solitary_mortality_risk/

[54] Solitary Watch. (2023, May). *Calculating torture.* https://solitarywatch.org/calculating-torture/

[55] Lantigua-Williams, J. (2016, December 5). *The link between race and solitary confinement.* The Atlantic. https://www.theatlantic.com/politics/archive/2016/12/race-solitary-confinement/509456/; The United States incarcerates a disproportionate number of people, compared to other countries. According to the American Civil Liberties Union (ACLU), Americans make up roughly 5% of the world's population but represent more than 20% of the world's prison population. Furthermore, there is a disproportionate number of Black and Hispanic males locked up in prison. The ACLU reports that, statistically speaking, 1 out of 3 Black boys and 1 out of 6 Latino boys is prison bound. This is compared to 1 out of 17 White boys.

[56] Prasow, A. (2020, December 14). *US architects join movement against capital punishment, torture.* Human Rights Watch. https://www.hrw.org/news/2020/12/14/us-architects-join-movement-against-capital-punishment-torture#:~:text=On%20December%2011%2C%20the%20Board,of%20prisoners%20for%2022%20hours

[57] Design Against Crime Research Centre. (2019). *Cell furniture: designs for safety, wellbeing, and sustainability.* https://files.cargocollective.com/c1114017/Cell-Furniture-Catalogue-_FINAL_compressed.pdf

[58] Rousseau J.-J. & Cress D. A. (1987). *Basic political writings.* Hackett Publishing Company.

[59] Mandela, N. (2009). *Long walk to freedom.* New York: Flash Point/Roaring Brook Press.

[60] Gun Violence Archive. (n.d.). *GVA 10 year review.* https://www.gunviolencearchive.org/

[61] Goldstick, J., Cunningham, R., & Carter, P. (2022). Current causes of death in children and adolescents in the United States. *New England Journal of Medicine, 20.* 1955-1956. https://www.nejm.org/doi/full/10.1056/nejmc2201761

[62] Latour, B. (2000). *Pandora's hope: essays on the reality of science studies.* Harvard University Press.

[63] Monteiro, M. (2019). *Ruined by design: How designers destroyed the world, and what we can do to fix it.* Independently published.

[64] Fisher, T. & Gamman, L. (2019). *Tricky design: the ethics of things.* Bloomsbury Visual Arts.

[65] Selinger, E. (2012, July 23). *The philosophy of the technology of the gun.* The Atlantic. https://www.theatlantic.com/technology/archive/2012/07/the-philosophy-of-the-technology-of-the-gun/260220/

[66] Reyes, P. (n.d.). *Palas por pistolas.* http://pedroreyes.net/08.php?szLang=en&Area=work&SubArea=08

[67] RAWTools. (n.d.). *Forging peace. Disarming hearts.* https://rawtools.org/

[68] Everytown For Gun Safety. (2023, December 15). U.S. Surpasses 2022 Total Unintentional Shootings by Children. https://everytownsupportfund.org/press/u-s-surpasses-2022-total-unintentional-shootings-by-children/

Chapter 6

[1] Though this is not the focus of this chapter, it would be worthwhile for the reader to review Peter Singer's views on abortion and end-of-life care for humans with severe disabilities and mental incapacitation, which demonstrate how he counters the pro-life view that all human life is sacrosanct.

[2] Singer, P. (2009). *Animal liberation; The definitive classic of the animal movement.* Open Road Integrated Media, Inc.

[3] PETA. (n.d.). *Our mission statement.* https://www.peta.org/about-peta/

[4] PETA. (n.d.). *Why animal rights?* https://www.peta.org/about-peta/why-peta/why-animal-rights/

[5] Moore, T. (2019, July17). *Inside the design evolution of the modern zoo.* Bloomberg. https://www.bloomberg.com/news/articles/2019-07-17/the-history-of-humans-looking-at-animals-in-zoos

[6] Nolen, S. (2002, November 5). *Designing zoo habitats that promote animal well-being.* American Veterinary Medical Association. https://www.avma.org/javma-news/2002-12-01/designing-zoo-habitats-promote-animal-well-being#:~:text=Research%20has%20also%20shown%20that,for%20and%20catch%20their%20food.

[7] Fekete, O.B., & Szilágyi, K.M. (2021) Contemporary zoo design and the historical treasure of garden art: Tiergarten Schönbrunn. *Journal of Landscape Architecture, 16,* 82-93, DOI: 10.1080/18626033.2021.2046809

[8] PETA. (n.d.). *How to tell if a place is a REAL animal sanctuary.* https://www.peta.org/features/real-animal-sanctuary-zoo/

[9] Roe, K., McConney, A., & Mansfield, C. F. (2014). How do zoos 'talk' to their general visitors? Do visitors 'listen'? A mixed method investigation of the communication between modern zoos and their general visitors. *Australian Journal of Environmental Education, 30*(2), 167–186. https://www.jstor.org/stable/26422865

[10] Fiby, M. (2008). Trends in zoo design – changing needs in keeping wild animals for a visiting audience. *The International Review of Landscape Architecture and Urban Design.* https://zoolex.org/page/publications/fiby/trends_in_zoodesign

[11] World Association of Zoos and Aquariums (2016). Future of zoo and aquarium design. *WAZA Magazine,* 17. https://www.waza.org/wp-content/uploads/2019/02/Gusset__Chin_2016.pdf

[12] World Association of Zoos and Aquariums (2016). Future of zoo and aquarium design. *WAZA Magazine,* 17. https://www.waza.org/wp-content/uploads/2019/02/Gusset__Chin_2016.pdf

[13] Singer, P. (2011, July 13). *Moral progress and animal welfare.* Project Syndicate. https://www.project-syndicate.org/commentary/moral-progress-and-animal-welfare#:~:text=PRINCETON%20%E2%80%93%20Mahatma%20Gandhi%20acutely%20observed,mark%20of%20a%20civilized%20society.

[14] Gandhi, M. (2018). *An autobiography or the story of my experiments with truth: a critical edition.* New Haven: Yale University Press. https://doi.org/10.12987/9780300238402
[15] *Young India:*(1919-1932) English weekly journal, published from Bombay as a bi-weekly, under Gandhiji's supervision from May 7, 1919, and as a weekly from Ahmedabad, with Gandhiji as editor from October 8, 1919.
[16] PETA. (n.d.). *Don't zoos help to preserve endangered species?* https://www.peta.org/about-peta/faq/dont-zoos-help-to-preserve-endangered-species/
[17] ASPCA. (n.d.). *Position statement on zoos and aquariums.* https://www.aspca.org/about-us/aspca-policy-and-position-statements/position-statement-zoos-and-aquariums
[18] Rees, P. A. (2023). *Zoo studies: living collections, their animals, and visitors.* Cambridge: Cambridge University Press.
[19] Vander Zanden, M.J. (2005). The success of animal invaders. *Proc Natl Acad Sci USA, 102,* 7055–7056.
[20] Bolotnikova, M. (2021, November 28). It's time to stop demonizing "invasive" species. *Vox.* https://www.vox.com/down-to-earth/22796160/invasive-species-climate-change-range-shifting
[21] The 2023 first and second place winners captured and killed 20 and 14 Burmese pythons, respectively.
[22] The Nature Conservancy. (2019, October 14). *Python patrol: stopping a Burmese python invasion.* https://www.nature.org/en-us/about-us/where-we-work/united-states/florida/stories-in-florida/stopping-a-burmese-python-invasion/?gad_source=1&gclid=CjwKCAiA-vOsBhAAEiwAIWR0TYgB3jGDGjnqUgErsxm5N-venWNH-hEP5ZLeDM9N_8PDIOPpTWSggRoCUUsQAvD_BwE&gclsrc=aw.ds
[23] Invasive Species Advisory Committee. (2016, December 6). *Invasive species impacts on infrastructure.* https://www.doi.gov/sites/doi.gov/files/uploads/isac_infrastructure_white_paper.pdf
[24] Keim, B. (2018, August 22). *Is New Zealand's war on invasive predators an expensive distraction?* Anthropocene. https://www.anthropocenemagazine.org/2018/08/new-zealand-predator-free-critique/?gclid=CjwKCAiA-vOsBhAAEiwAIWR0TQjDGM5UcWUN6Dh3FmechhPgGrMMbJsKt0bcMIOPteimD-W931LQeBoCMGcQAvD_BwE
[25] Novak B. J. (2018). De-Extinction. *Genes, 9*(11), 548. https://doi.org/10.3390/genes9110548
[26] Browning, H. (2018). Won't somebody please think of the mammoths? De-extinction and animal welfare. *Journal of Agricultural and Environmental Ethics. 31*(6): 785-803
[27] Newcomb, T. (2023, January 30). *Scientists are reincarnating the woolly mammoth to return in 4 years.* Popular Mechanics. https://www.popularmechanics.com/science/animals/a42708517/scientists-reincarnating-woolly-mammoth/
[28] Colossal Bioscience. (2022, May 4). *Woolly mammoth de-extinction: restoring the past for a better future* [Video]. YouTube. https://www.youtube.com/watch?v=wUtl7kVP16c&t=67s
[29] Odenbaugh, J. (2023). Philosophy and ethics of de-extinction. *Cambridge Prisms: Extinction, 1*(7). doi:10.1017/ext.2023.4

[30] Martindale, D. (2023, April 26). *Bringing back woolly mammoths and dodos is a bad idea.* Vox. https://www.vox.com/future-perfect/23696294/de-extinction-colossal-biosciences-woolly-mammoth-dodo-ethics
[31] Williams, E. E. & DeMello, M. (2007). *Why animals matter: the case for animal protection.* Prometheus Books.
[32] Williams, E. E. & DeMello, M. (2007). *Why animals matter: the case for animal protection.* Prometheus Books.
[33] Ufer, D. (2023, April 24). *Farm animal welfare policies cover breeding sows, veal calves, or laying hens in 14 U.S. states.* USDA Economic Research Service. https://www.ers.usda.gov/amber-waves/2023/april/farm-animal-welfare-policies-cover-breeding-sows-veal-calves-or-laying-hens-in-14-u-s-states/
[34] The Humane League. (2020, December 3). *Everything you should know about battery cages.* https://thehumaneleague.org/article/battery-cages
[35] Besl, J. (2001). A short retrospective on 20th century U.S. population change. *Indiana Business Review.* https://www.ibrc.indiana.edu/ibr/2001/spring01/03.pdf
[36] Davis, C. et al. (2022, August). U.S. hog production: rising output and changing trends in productivity growth. *USDA Economic Research Service.* https://www.ers.usda.gov/webdocs/publications/104437/err-308.pdf?v=6674.2
[37] Sinclair, U. (1906). *The jungle.* New York, New American Library.
[38] Torrella, K. (2023, December 12). *We raise 18 billion animals a year to die – and then we don't even eat them.* Vox. https://www.vox.com/future-perfect/22890292/food-waste-meat-dairy-eggs-milk-animal-welfare
[39] Dalzell, N. (2021, February 10). *Dangerous working conditions on factory farms.* New Roots Institute. https://ffacoalition.org/articles/dangerous-conditions-factory-farms/
[40] Wasley, A. (2018, July 5). *Two amputations a week: the cost of working in a US meat plant.* The Guardian. https://www.theguardian.com/environment/2018/jul/05/amputations-serious-injuries-us-meat-industry-plant
[41] Dalzell, N. (2021, February 10). *Dangerous working conditions on factory farms.* New Roots Institute. https://ffacoalition.org/articles/dangerous-conditions-factory-farms/
[42] Food Print. (2023, February 28). *How industrial agriculture affects our air.* https://foodprint.org/issues/how-industrial-agriculture-affects-our-air/
[43] Here are some websites that locate such farms: https://www.americanhumane.org/humane-heartland/, https://cratefreeusa.org/find-humane/, https://www.farmerfocus.com/certified-humane/
[44] As an April Fools' Day prank, PETA even posted a press release announcing their partnership with Crayola and their new product, the Little Animal Rights Activist's First Paint-Throwing Kit, which includes a water gun for "shooting an arc of water-based, washable paint on anyone wearing wool, fur, leather, or any other cruelly obtained animal-derived clothing!" Crayola did not find the joke to be funny.
[45] Holland, O. (2020, February 5). *PETA ends "I'd rather go naked" anti-fur campaign after three decades.* CNN. https://www.cnn.com/style/article/peta-naked-fur-campaign-ends/index.html

[46] Alexander, E. (2014, July 2). Joan Rivers responds to Peta protestors: 'Those furs that would have been lying in a cellar have gotten to go to the opera.' *Independent.* https://www.independent.co.uk/news/people/joan-rivers-responds-to-peta-protesters-those-furs-that-would-have-been-lying-in-a-cellar-have-gotten-to-go-to-the-opera-9579978.html

[47] Thomas, S. (2011). Situated empathy: constructed theoretical discourse addressing the empathetic motivations shared by fashion design for sustainability, and the potential of socially engaged Buddhist ethics to inform design practice. Ph.D. Royal Melbourne Institute of Technology University. https://s3-ap-southeast-1.amazonaws.com/ap-st01.ext.exlibrisgroup.com/61RMIT_INST/storage/alma/78/9C/66/BF/AE/B1/8B/46/60/D5/1B/FC/A2/E1/7B/9E/Thomas.pdf?response-content-type=application%2Fpdf&X-Amz-Algorithm=AWS4-HMAC-SHA256&X-Amz-Date=20240119T185140Z&X-Amz-SignedHeaders=host&X-Amz-Expires=119&X-Amz-Credential=AKIAJN6NPMNGJALPPWAQ%2F20240119%2Fap-southeast-1%2Fs3%2Faws4_request&X-AmzSignature=436f24d4410537aad5abfbcaf71cbea25c3a626a408dbff9a790b58345763cea

[48] Thomas, S. (2018). *Fashion ethics.* Routledge. https://doi.org/10.4324/9781315109367

Chapter 7

[1] Thoreau, H. D. (1917) *Walden, or, Life in the woods.* Chicago, New York: Scott, Foresman and Company

[2] Carr, N. (2008, July/August). Is Google making us stupid? *The Atlantic.* https://www.theatlantic.com/magazine/archive/2008/07/is-google-making-us-stupid/306868/

[3] Zinsser, W. (2006). *On writing well (30th ed.).* HarperCollins.

[4] Carr, N. (2011). *The shallows.* WW Norton.

[5] Hopkins, P.D. (2012). Transhumanism. Encyclopedia of Applied Ethics (Second Edition). 414-422. https://doi.org/10.1016/B978-0-12-373932-2.00243-X

[6] O'Connell, M. (2017). *To be a machine: adventures among cyborgs, utopians, hackers, and the futurists solving the modest problem of death.* Doubleday.

[7] Levin, N. (2019). Introduction to Philosophy and the Ship of Theseus. In *Ancient Philosophy Reader, An Open Educational Resource.* NGE Far Press. https://human.libretexts.org/Bookshelves/Philosophy/Book%3A_Ancient_Philosophy_Reader_(Levin)/01%3A_The_Start_of_Western_Philosophy_and_the_Pre-Socratics/1.01%3A_Introduction_to_Philosophy_and_the_Ship_of_Theseus

[8] Galton, F. (1865). Hereditary talent and character. *Macmillan's Magazine, 12,* 157-166.

[9] Cold Spring Harbor Laboratory DNA Learning Center. (n.d.). *T. Roosevelt letter to C. Davenport about "degenerates reproducing".* https://dnalc.cshl.edu/view/11219-T-Roosevelt-letter-to-C-Davenport-about-degenerates-reproducing-.html

[10] Nielsen, K. (2004). *The radical lives of Helen Keller.* New York University Press.

[11] It should be noted that, in recent years, there has been some debate over the originality of Crick and Watson's research into discovering the double helix of DNA. Although the two men are credited as the "fathers of DNA", new claims have surfaced that their work echoed the research of others, including Rosalind Franklin and Maurice Wilkins, who were never recognized formally for their contributions. As is often the case, new research

scaffolds on to previous research, but in this case, the gender discrimination most likely played a role, as it did throughout the rest of 1950s British society.

[12] Watson, J. & Crick, F. (1953). Molecular structure of nucleic acids: a structure for deoxyribose nucleic acid. *Nature,* 171, 737–738.

[13] Science. (1997, May 6). *Watson urges "put Hitler behind us."* https://www.science.org/content/article/watson-urges-put-hitler-behind-us

[14] Wellcome. (2003, March 30). *Francis Crick's controversial archive on first public display.* https://wellcome.org/press-release/francis-cricks-controversial-archive-first-public-display

[15] Doudna, J. & Sternberg, S. (2017). *A crack in creation: gene editing and the unthinkable power to control evolution.* Houghton Mifflin Harcourt.

[16] Metzl, J. F. (2019). *Hacking Darwin: genetic engineering and the future of humanity.* Sourcebooks, Inc.

[17] Carlson, T. (2020). *Ship of fools: how a selfish ruling class is bringing America to the brink of revolution.* Free Press.

[18] Rainie, L., Funk, C., Anderson, M., & Tyson, A. (2022, March 17). *Americans are closely divided over editing a baby's genes to reduce serious health risk.* Pew Research Center. https://www.pewresearch.org/internet/2022/03/17/americans-are-closely-divided-over-editing-a-babys-genes-to-reduce-serious-health-risk/

[19] Science. (1997, May 6). *Watson urges "put Hitler behind us."* https://www.science.org/content/article/watson-urges-put-hitler-behind-us

[20] Reuters. (2023, September 12). *US-Mexico border is world's deadliest land migration route, IOM finds.* https://www.reuters.com/world/us-mexico-border-is-worlds-deadliest-land-migration-route-iom-finds-2023-09-12/

[21] Spagat, E. (2023, December 19). *Illegal crossings surge in remote areas as Congress and the White House weigh major asylum limits.* AP News. https://apnews.com/article/arizona-border-wall-breaches-asylum-immigration-beddd2ca0ffd02884ce0d0ad1785f538

[22] Pew Research Center. (2022, March 15). *Public expresses mixed views of U.S. response to Russia's invasion of Ukraine.* https://www.pewresearch.org/politics/2022/03/15/public-expresses-mixed-views-of-u-s-response-to-russias-invasion-of-ukraine/

[23] Richards, Z. (2023, December 19). *Trump doubles down on immigrant 'blood' remark, says he 'never read Mein Kampf.'* NBC News. https://www.nbcnews.com/politics/donald-trump/trump-doubles-immigrant-blood-remark-says-never-read-mein-kampf-rcna130535

[24] Portes, J. (2019). The Economics of Migration. *Contexts, 18*(2), 12-17. https://doi.org/10.1177/1536504219854712

[25] The Intercept. (n.d.). *Border Patrol Strategic Plan 1994 and Beyond.* https://www.documentcloud.org/documents/5987025-Border-Patrol-Strategic-Plan-1994-and-Beyond.html

[26] President Joe Biden campaigned on the promise that he would not build "another foot of wall" and even halted construction efforts on his first day in office, yet his administration has appropriated funds to add more than 20 miles of wall along the Rio Grande valley in Texas.

[27] Rascoe, A. & Kocherga, A. (2023, October 8). *The Biden administration is building a controversial part of the border wall in Texas.* NPR News. https://www.npr.org/2023/10/08/1204545880/the-biden-administration-is-building-a-controversial-part-of-the-border-wall-in-

[28] Sinclair, W. (1978, December 23). *U.S. and Mexico embroiled in dispute over 'tortilla curtain.'* The Washington Post. https://www.washingtonpost.com/archive/politics/1978/12/24/us-and-mexico-embroiled-in-dispute-over-tortilla-curtain/0f9b055a-5989-4600-8ecb-a642b8a14777/

[29] The White House. (2023, October 5). *Remarks by President Biden before national security team briefing on Ukraine.* https://www.whitehouse.gov/briefing-room/speeches-remarks/2023/10/05/remarks-by-president-biden-before-national-security-team-briefing-on-ukraine/

[30] Overmyer-Velazquez, M. (2011). *Beyond La Frontera: the history of Mexico-U.S. migration,* Oxford University Press.

[31] The Intercept. (n.d.). *Border Patrol Strategic Plan 1994 and Beyond.* https://www.documentcloud.org/documents/5987025-Border-Patrol-Strategic-Plan-1994-and-Beyond.html

[32] Rosenberg, M., Cooke, K., & Trotta, D. (2022, July 25). *The border's toll.* Reuters. https://www.reuters.com/investigates/special-report/usa-immigration-border-deaths/

[33] Rivera, N. (2020). *Painted over in brown: border art as visual discourse of resistance.* UNM Digital Repository. https://digitalrepository.unm.edu/cgi/viewcontent.cgi?article=1140&context=cj_etds

[34] Liepert, A.E., Berndtson, A.E., & Hill, L.L. (2022). Association of 30-ft US-Mexico border wall in San Diego with increased migrant deaths, trauma center admissions, and injury severity. *JAMA Surg, 157*(7), 633–635.

[35] Agid, S. (2019). Making safety, making freedom: problem-setting, collaborative design, and contested futures. In T. Fisher and L. Gamman (Eds.), *Tricky design: the ethics of things.* Bloomsbury Visual Arts.

[36] Schön, D. A. (1983). *The reflective practitioner: how professionals think in action.* New York: Basic Books.

[37] Agid, S. (2019). Making safety, making freedom: problem-setting, collaborative design, and contested futures. In T. Fisher and L. Gamman (Eds.), *Tricky design: the ethics of things.* Bloomsbury Visual Arts.

[38] Grant, J. & Fox, F. (1992). Understanding the role of the designer in society. *Journal of Art & Design Education, 11,* 77-87. https://doi.org/10.1111/j.1476-8070.1992.tb00689.x

[39] Bennett, P. (2014, September 2). Can death be designed? *Medium.* https://medium.com/ideo-stories/can-death-be-designed-c002f4bcb383

[40] Nilson, E. et al. (2022). Premature deaths attributable to the consumption of ultraprocessed foods in Brazil. *Global Health Promotion and Prevention, 64*(1), 129-136.

[41] Berman, R. (2018, March 16). Startup offers a path to immortality. The catch? It's 100% fatal. *Big Think.* https://bigthink.com/surprising-science/company-offers-a-killer-new-way-to-upload-your-mind/

[42] Swan, R. (2024, January 6). *The Golden Gate Bridge suicide nets are nearly finished. The milestone comes with warnings from the future.* San Francisco Chronicle. https://www.sfchronicle.com/sf/article/golden-gate-bridge-suicide-18589579.php
[43] Laranjeira, C., Dixe, M. A., Querido, A., & Stritch, J. M. (2022). Death cafés as a strategy to foster compassionate communities: Contributions for death and grief literacy. *Frontiers in psychology, 13,* 986031. https://doi.org/10.3389/fpsyg.2022.986031
[44] The Death Designer. (n.d.). *Designing a better deathcare for everyone.* https://www.thedeathdesigner.com/
[45] Dignitas. (n.d.). *Who is Dignitas?* http://www.dignitas.ch/?lang=en
[46] A Sacred Passing. (n.d.). *A sacred passing: death midwifery & community education.* Givelynow. https://givelynow.com/o/a-sacred-passing
[47] Fisher, T. & Gamman, L. (2019). *Tricky design: the ethics of things.* Bloomsbury Visual Arts.
[48] Casti, J. (1998). "Byteing" back. *Complexity.* https://onlinelibrary.wiley.com/doi/pdf/10.1002/%28SICI%291099-0526%28199801/02%293%3A3%3C8%3A%3AAID-CPLX2%3E3.0.CO%3B2-6
[49] Applebaum, J., MacLean, E., & McDonald, S. (2021). Love, fear, and the human-animal bond: on adversity and multispecies relationships. *Comprehensive Psychoneuroendocrinology, 7.* https://doi.org/10.1016/j.cpnec.2021.100071
[50] Herold, E. (2024). *Robots and the people who love them: holding on to our humanity in an age of social robots.* First edition. New York, St. Martin's Press.
[51] Szondy, M. & Fazekas, P. (2024). Attachment to robots and therapeutic efficiency in mental health. *Frontiers in Psychology, 15.* https://www.frontiersin.org/journals/psychology/articles/10.3389/fpsyg.2024.1347177/full

Chapter 8

[1] Goodman, P. (1969). "Can Technology Be Humane?" *New York Review of Books, 13*(9).
[2] Human Rights Watch. (2023, February 14). *US: New policy on autonomous weapons flawed.* https://www.hrw.org/news/2023/02/14/us-new-policy-autonomous-weapons-flawed?gad_source=1&gclid=CjwKCAiAopuvBhBCEiwAm8jaMR32yi1cMepw2JbYWekty8_ITP69BbC0PHbnJ-3H6i8Z7UmBam2aCBoC6zsQAvD_BwE
[3] To be clear, there is a difference between being *humane* and being *ethical,* though they are closely related. Humane refers to the compassionate treatment of others; ethics refers to mindful wellbeing. One cannot be ethical without also being humane, but humane behavior can be unethical. For instance, a recreational hunter can kill an animal humanely, but this does not make the act of recreational hunting ethical.
[4] Postman, N. (1993). *Technopoly: the surrender of culture to technology.* 1st Vintage Books ed. New York, Vintage Books.
[5] Postman, N. (1999). *Building a bridge to the 18th century: how the past can improve our future.* Alfred A. Knopf: Distributed by Random House.
[6] Postman, N. (1993). *Technopoly: the surrender of culture to technology.* 1st Vintage Books ed. New York, Vintage Books.

[7] Palm, E. & Hansson, S.O. (2006). The case for ethical technology assessment (eTA). *Technological Forecasting and Social Change, 73*(5), 543-558.
[8] Markkula Center for Applied Ethics. (2018, June 22). *An ethical toolkit for engineering/design practice.* https://www.scu.edu/ethics-in-technology-practice/ethical-toolkit/
[9] Thierer, A. (2024, October). *Defining "Technology."* Medium. https://medium.com/@AdamThierer/defining-technology-0c5c95c8360d
[10] Gigante, M. (1993). Virtual reality: definitions, history, and applications. *Virtual Reality Systems,* 3-14.
[11] Lanier, J. (2017). *Dawn of the new everything: encounters with reality and virtual reality.* Henry Holt and Company.
[12] I wrote a book in 2018, *The Smartphone Paradox,* that addresses this digital phenomenon and its devastating consequences on our social, emotional, cognitive, and behavioral health.
[13] Seife, C. (2014). *Virtual unreality.* Penguin Group.
[14] Seife, C. (2014). *Virtual unreality.* Penguin Group.
[15] Hoffman, H. et al., (2022). Virtual reality as an adjunctive non-pharmacologic analgesic for acute burn pain during medical procedures. *Annals of Behavioral Medicine, 4*(2), 183-191.
[16] Ramirez, E. J. (2022). *The ethics of virtual and augmented reality: Building worlds.* Routledge.
[17] Slater, M. (2020). The ethics of realism in virtual and augmented reality. *Frontiers in Virtual Reality, 1*(1). doi: 10.3389/frvir.2020.00001
[18] Luckey, P. (2022, November 6). *If you die in the game, you die in real life.* The Blog of Palmer Luckey. https://palmerluckey.com/if-you-die-in-the-game-you-die-in-real-life/
[19] Narcotta, J. & Ablondi, B. (2022, June 22). *Strategy analytics: Amazon's Ring remained atop the video doorbell market in 2021.* Business Wire. https://www.businesswire.com/news/home/20220622005023/en/Strategy-Analytics-Amazons-Ring-Remained-atop-the-Video-Doorbell-Market-in-2021#:~:text=These%20five%20companies%20combined%20to,to%20consumers%20globally%20in%202021.
[20] Rainie, L., Funk, C., Anderson, M., & Tyson, A. (2022, March 17). *Public more likely to see facial recognition use be police good, rather than bad for society.* Pew Research Center. https://www.pewresearch.org/internet/2022/03/17/public-more-likely-to-see-facial-recognition-use-by-police-as-good-rather-than-bad-for-society/
[21] Georgetown Law Center on Privacy & Technology. (2016, October 18). *The perpetual line-up.* https://www.perpetuallineup.org/findings/racial-bias
[22] Gender Shades Project. (n.d.). *Gender Shades.* http://gendershades.org/overview.html
[23] Fry, H. (2018). *Hello world: being human in the age of algorithms.* W.W. Norton & Company.
[24] Stanford Center on China's Economy and Institutions. (2022, September 15). *Who supports social credit systems in China?* https://sccei.fsi.stanford.edu/china-briefs/who-supports-social-credit-systems-china

[25] Smart Security Pros. (2021, March 19). *Keep schools safe with advanced video surveillance.* https://smartsecuritypros.com/keep-schools-safe-with-video-surveillance/
[26] Imagine humane technology. (2021, August 17). *Center for Humane Technology.* https://www.humanetech.com/youth/imagine-humane-technology
[27] Ahmed, S. K. (2023). Road traffic accidental injuries and deaths: A neglected global health issue. *Health Science Reports, 6*(5).
[28] Carr, N. (2014). *The glass cage: how our computers are changing us.* W.W. Norton & Company.
[29] Merchant, B. (2023). *Blood in the machine: the origins of the rebellion against big tech.* Little, Brown, and Company.
[30] Deichmann, J., Ebel, E., Heineke, K., Heuss, R., Kellner, M., & Steiner, F. (2023, January 6). *Autonomous driving's future: Convenient and connected.* McKinsey & Company. https://www.mckinsey.com/industries/automotive-and-assembly/our-insights/autonomous-drivings-future-convenient-and-connected
[31] Waymo. (n.d.). *The world's most experienced driver.* https://waymo.com/
[32] Coyne, C. J., & Hall, A. R. (2018). The drone paradox: fighting terrorism with mechanized terror. *The Independent Review, 23*(1), 51–67. http://www.jstor.org/stable/26591799
[33] Phelps, W. (2021). *On killing remotely: the psychology of killing with drones.* Little, Brown, and Company.
[34] Khurshudyan, I., Ilushina, M., & Khudov, K. (2022, December 2). *Russia and Ukraine are fighting the first full-scale drone war.* The Washington Post. https://www.washingtonpost.com/world/2022/12/02/drones-russia-ukraine-air-war/
[35] Singer, P. (2009, May 18). *Isaac Asimov's laws of robotics are wrong.* Brookings. https://www.brookings.edu/articles/isaac-asimovs-laws-of-robotics-are-wrong/

Chapter 9

[1] Monteiro, M. (2019). *Ruined by design: How designers destroyed the world, and what we can do to fix it.* Independently published.
[2] Thunberg, G. (2019). *Full speech to world leaders at UN Climate Action Summit.* [Video]. YouTube. https://www.youtube.com/watch?v=KAJsdgTPJpU
[3] Tyson, A. & Kennedy, B. (2023, October 25). *How Americans view future harms from climate change in their community and around the U.S.* Pew Research Center. https://www.pewresearch.org/science/2023/10/25/how-americans-view-future-harms-from-climate-change-in-their-community-and-around-the-u-s/
[4] For a complete list of the UN's Sustainable Development Goals, refer to Chapter 4.
[5] Winner L. (1986). *The whale and the reactor: a search for limits in an age of high technology.* University of Chicago Press.
[6] Hippo Roller. (n.d.). *Revolutionizing water access: the resurgence of Hippo Roller technology.* https://hipporoller.org
[7] Verbeek, P.-P. (2011). *Moralizing technology: understanding and designing the morality of things.* The University of Chicago Press.
[8] Papanek, V. (1998). Towards the spiritual in design. In Ostrow, S. (Ed.), *Beauty is nowhere: ethical issues in art and design.* Routledge. https://doi.org/10.4324/9780203871423

[9] Carr, N. (2014). The glass cage: how our computers are changing us. W.W. Norton & Company.

References

A Sacred Passing. (n.d.). *A sacred passing: Death midwifery & community education.* Givelynow. https://givelynow.com/o/a-sacred-passing

Aagaard, J. (2021). The passive subject: a phenomenological contribution to STS. *Social Epistemology Review and Reply Collective, 10*(5), 14-19. https://wp.me/p1Bfg0-5Pb

ABC News. (2008, June 5). *Car bait program goes awry as thief steals car, kills woman.* ABC news. https://abcnews.go.com/GMA/story?id=5004251&page=1

ACLU. (2005, November 23). *Racial profiling: definition.* ACLU. https://www.aclu.org/documents/racial-profiling-definition

ACLU. (2023, August 6). *After third wrongful arrest, ACLU slams Detroit police department for continuing to use faulty facial recognition technology.* ACLU. https://www.aclu.org/press-releases/after-third-wrongful-arrest-aclu-slams-detroit-police-department-for-continuing-to-use-faulty-facial-recognition-technology

ACLUSoCal. (2022, October). *Reimagining community safety in California.* https://catalyst-ca.cdn.prismic.io/catalyst-ca/126c30a8-852c-416a-b8a7-55a90c77a04e_APCA+ACLU+REIMAGINING+COMMUNITY+SAFETY+2022_5.pdf

Agid, S. (2019). Making safety, making freedom: problem-setting, collaborative design, and contested futures. In T. Fisher and L. Gamman (Eds.), *Tricky design: the ethics of things.* Bloomsbury Visual Arts. https://doi.org/10.5040/9781474277211.ch-008

Ahmed, S. K. (2023). Road traffic accidental injuries and deaths: A neglected global health issue. *Health Science Reports, 6*(5). https://doi.org/10.1002/hsr2.1240

AIA. (2023, November 29). *Climate action plan.* AIA. https://www.aia.org/resources/6307290-climate-action-plan

Akkad, D. (2022, October 7). Neom: Saudi Arabia sentences tribesmen to death for resisting displacement. *Middle East Eye.* https://www.middleeasteye.net/news/neom-saudi-arabia-sentences-tribesmen-death-resisting-displacement

Alexander, C. (1979). *The timeless way of building.* Oxford University Press, New York.

Alexander, E. (2014, July 2). Joan Rivers responds to Peta protestors: 'Those furs that would have been lying in a cellar have gotten to go to the opera.' *Independent.* https://www.independent.co.uk/news/people/joan-rivers-responds-to-peta-protesters-those-furs-that-would-have-been-lying-in-a-cellar-have-gotten-to-go-to-the-opera-9579978.html

American Museum of Natural History. (n.d.). *The Manhattan Project.* https://www.amnh.org/exhibitions/einstein/peace-and-war/the-manhattan-project

Anthes, E. (2020). *The great indoors: the surprising science of how buildings shape our behavior, health, and happiness* (First Picador paperback edition). Scientific American/Picador.

Apple. (2021, April 20). *Apple introduces AirTag.* Apple. https://www.apple.com/newsroom/2021/04/apple-introduces-airtag/

Applebaum, J., MacLean, E., & McDonald, S. (2021). Love, fear, and the human-animal bond: on adversity and multispecies relationships. *Comprehensive Psychoneuroendocrinology, 7.* https://doi.org/10.1016/j.cpnec.2021.100071

Architecture 2030. (n.d.). *Why the built environment?* https://www.architecture2030.org/why-the-built-environment/

Ariely, D. (2012). *The (Honest) Truth About Dishonesty.* HarperCollins.

Artemis. (n.d.). *The Artemis missions.* NASA. https://www.nasa.gov/specials/artemis/#why

ASPCA. (n.d.). *Position statement on zoos and aquariums.* https://www.aspca.org/about-us/aspca-policy-and-position-statements/position-statement-zoos-and-aquariums

AtomicArchive.com. (n.d.). *Einstein's letter to President Roosevelt – 1939.* https://www.atomicarchive.com/resources/documents/beginnings/einstein.html

AtomicArchive.com. (n.d.). *Report of the committee on political and social problems.* https://www.atomicarchive.com/resources/documents/manhattan-project/franck-report.html#V

Axon, S. (2021, November 23). Bluetooth tracking company Tile acquired for $205 million. *Ars Technica.* https://arstechnica.com/gadgets/2021/11/bluetooth-tracking-company-tile-acquired-for-205-million/. https://doi.org/10.1038/s41415-021-3380-6

Baba, A. (2021, November 9). *Buttigieg is right – America's roads are racist. Ted Cruz and Tucker Carlson are pretending to misunderstand.* Independent. https://www.independent.co.uk/voices/buttigieg-racist-ted-cruz-tucker-carlson-b1954576.html

Baker, J. O., Bader, C. D., & Hirsch, K. (2015). Desecration, Moral Boundaries, and the Movement of Law: The Case of Westboro Baptist Church. *Deviant Behavior, 36*(1), 42–67. https://doi.org/10.1080/01639625.2014.906282

Bakewell, S. (2013, November 22). *Clang went the trolley.* The New York Times. https://www.nytimes.com/2013/11/24/books/review/would-you-kill-the-fat-man-and-the-trolley-problem.html

Balkan, A. & Kalbag, L. (n.d.). *Ethical design.* INDIE. https://ind.ie/ethical-design/

Barlow, A. (n.d.). *Architecture styles in North Carolina courthouses.* North Carolina Administrative Office of the Courts. https://www.nccourts.gov/assets/documents

/publications/Brief-History-Architecture-NC-Courthouses.pdf?XMxou7Lols3FqbC5XrzqG.LpWZocCQZR

Barr, S. (2007). Factors influencing environmental attitudes and behaviors: a U.K. case study of household waste management. *Environment and Behavior, 39*(4), 435-473. https://doi.org/10.1177/0013916505283421

Bayor, R. H. (2000). *Race and the Shaping of Twentieth-Century Atlanta.* University of North Carolina Press.

Beck, B., Holder, E., Novak, A. & Kaplan, J. (2023). The material of policing: budgets, personnel, and the United States' misdemeanor arrest decline. *The British Journal of Criminology, 63*(2). 330-347. https://academic.oup.com/bjc/article-abstract/63/2/330/6568129?redirectedFrom=fulltext&login=false. https://doi.org/10.1093/bjc/azac005

Bejan, A. & Zane, J. P. (2012). *Design in nature.* Knopf Doubleday Publishing Group. https://doi.org/10.1115/1.2012-JUN-4

Bejan, A., & Lorente, S. (2010). The constructal law of design and evolution in nature. *Philosophical transactions of the Royal Society of London. Biological sciences, 365*(1545), 1335–1347. https://doi.org/10.1098/rstb.2009.0302

Benjamin, R. (2012, March 29). *The gated community mentality.* The New York Times. https://www.nytimes.com/2012/03/30/opinion/the-gated-community-mentality.html

Benko, J. (2015, March 26). *The radical humaneness of Norway's Halden Prison.* The New York Times. https://www.nytimes.com/2015/03/29/magazine/the-radical-humaneness-of-norways-halden-prison.html

Bennett, P. (2014, September 2). Can death be designed? *Medium.* https://medium.com/ideo-stories/can-death-be-designed-c002f4bcb383

Bentham, J. (1791). *Panopticon or the Inspection House.* Payne: London.

Berman, R. (2018, March 16). Startup offers a path to immortality. The catch? It's 100% fatal. *Big Think.* https://bigthink.com/surprising-science/company-offers-a-killer-new-way-to-upload-your-mind/

Besl, J. (2001). A short retrospective on 20th century U.S. population change. *Indiana Business Review.* https://www.ibrc.indiana.edu/ibr/2001/spring01/03.pdf

Bever, L. (2015, March 5). K-Cups inventor: I feel bad sometimes that I ever did it. *The Washington Post.* https://www.washingtonpost.com/news/morning-mix/wp/2015/03/05/k-cups-inventor-i-feel-bad-sometimes-that-i-ever-did-it/

Bevis, C. & J. Nutter (1978). *Changing street layouts to reduce residential burglary.* Minneapolis: Minnesota Crime Prevention Center.

Bigelow, K. (1991). *Point Break.* Twentieth Century Fox.

Blakely, E. J. & Snyder, M. G. (n.d.). *Fortress America: gated communities in the United States. The New York Times.* https://archive.nytimes.com/www.nytimes.com/books/first/b/blakely-fortress.html

Boeck, A. (2023, June 27). *Why Saudi Arabia's "The Line" isn't a revolution in urban living. EurekAlert!* https://www.eurekalert.org/news-releases/993409

Bolotnikova, M. (2021, November 28). It's time to stop demonizing "invasive" species. *Vox.* https://www.vox.com/down-to-earth/22796160/invasive-species-climate-change-range-shifting

Booz, M. L. (1986). *The constitutionality of mandatory seat belt laws.* (Report # VHTRC 86-R41). Virginia Highway & Transportation Research Council. https://rosap.ntl.bts.gov/view/dot/19122

Browning, H. (2018). Won't somebody please think of the mammoths? De-extinction and animal welfare. *Journal of Agricultural and Environmental Ethics, 31*(6), 785-803. https://doi.org/10.1007/s10806-018-9755-2

Burton-Harris, V. & Mayor, P. (2020, June 24). *Wrongfully arrested because face recognition can't tell black people apart.* ACLU. https://www.aclu.org/news/privacy-technology/wrongfully-arrested-because-face-recognition-cant-tell-black-people-apart

Cardenas, F. (2023, August 4). *Why Messi's arrival doesn't mean guaranteed sellouts at Inter Miami's stadium.* The Athletic. https://theathletic.com/4748102/2023/08/04/inter-miami-ticket-prices-messi/

Carlson, T. (2020). *Ship of fools: how a selfish ruling class is bringing America to the brink of revolution.* Free Press.

Carnegie Science Earth and Planets Laboratory. (2023, October 16). *Scientists, philosophers identify nature's missing evolutionary law.* ScienceDaily. https://www.sciencedaily.com/releases/2023/10/231016163100.htm

Caro, R. A. (1974). *The power broker: Robert Moses and the fall of New York.* Knopf: New York.

Carr, N. (2008, July/August). *Is Google making us stupid?* The Atlantic. https://www.theatlantic.com/magazine/archive/2008/07/is-google-making-us-stupid/306868/

Carr, N. (2011). *The shallows.* WW Norton.

Carr, N. (2014). *The glass cage: how our computers are changing us.* W.W. Norton & Company.

Casella, J., Ridgeway, J., & Shourd, S. (2016). *Hell is a very small place: voices from solitary confinement.* The New Press.

Casti, J. (1998). "Byteing" back. *Complexity.* https://onlinelibrary.wiley.com/doi/pdf/10.1002/%28SICI%291099-0526%28199801/02%293%3A3%3C8%3A%3AAID-CPLX2%3E3.0.CO%3B2-6. https://doi.org/10.1002/(SICI)1099-0526(199805/06)3:5%3C11::AID-CPLX2%3E3.3.CO;2-3

Centers for Disease Control and Prevention. (2020, April 28). *Smoking during pregnancy.* Centers for Disease Control and Prevention. https://www.cdc.gov/tobacco/basic_information/health_effects/pregnancy/index.htm

Central Park Conservancy. (n.d.). *About us.* https://www.centralparknyc.org/about

Chakrabarty, P. (2023, January 24). The basics you need to know & understand about speculative design. *Medium.* https://uxplanet.org/the-basics-you-need-to-know-and-understand-about-speculative-design-8a8bf5be4162

Chelsea Clip. (n.d.). *How it works.* Chelsea Clip. https://www.chelseaclip.co.uk/pages/how-it-works

Chen, M. K., & Shapiro, J. M. (2007). Do harsher prison conditions reduce recidivism? A discontinuity-based approach. *American Law and Economics Review,* 9(1), 1–29. http://www.jstor.org/stable/42705508. https://doi.org/10.1093/aler/ahr007

Churchill, W. (1943, October). *House of Commons Rebuilding.* https://api.parliament.uk/historic-hansard/commons/1943/oct/28/house-of-commons-rebuilding

Coalition for the Homeless. (2023, February). *Basic facts about homelessness: New York City.* https://www.coalitionforthehomeless.org/basic-facts-about-homelessness-new-york-city/

Cold Spring Harbor Laboratory DNA Learning Center. (n.d.). *T. Roosevelt letter to C. Davenport about "degenerates reproducing."* https://dnalc.cshl.edu/view/11219-T-Roosevelt-letter-to-C-Davenport-about-degenerates-reproducing-.html

Collins, W. J., & Shester, K. L. (2013). Slum clearance and urban renewal in the United States. *American Economic Journal: Applied Economics, 5*(1), 239–273. http://www.jstor.org/stable/43189425. https://doi.org/10.1257/app.5.1.239

Colossal Bioscience. (2022, May 4). *Woolly mammoth de-extinction: restoring the past for a better future* [Video]. YouTube. https://www.youtube.com/watch?v=wUtl7kVP16c&t=67s

Cooley, A. (2016, July 1). *Design is key to making recycling work.* National Recreation and Park Association. https://www.nrpa.org/parks-recreation-magazine/2016/july/design-is-key-to-making-recycling-work/

Cowell, A. (2008, April 8). *Negligent driving killed Diana, jury finds.* The New York Times. https://www.nytimes.com/2008/04/08/world/europe/08diana.html

Coyne, C. J., & Hall, A. R. (2018). The drone paradox: fighting terrorism with mechanized terror. *The Independent Review, 23*(1), 51–67. http://www.jstor.org/stable/26591799

Critical design FAQ. (n.d.). *Reading Design.* https://www.readingdesign.org/critical-design-faq

Dahmus, J.B. (2014). Can efficiency improvements reduce resource consumption? A historical analysis of ten activities. *Journal of Industrial Ecology, 18*(6), 883–897. https://doi.org/10.1111/jiec.12110

Dalzell, N. (2021, February 10). *Dangerous working conditions on factory farms.* New Roots Institute. https://ffacoalition.org/articles/dangerous-conditions-factory-farms/

Davis, C. et al. (2022, August). *U.S. hog production: rising output and changing trends in productivity growth.* USDA Economic Research Service. https://www.ers.usda.gov/webdocs/publications/104437/err-308.pdf?v=6674.2

de Boer. (2021). Explaining multistability: postphenomenology and affordances of technologies. *AI & Society.* https://doi.org/10.1007/s00146-021-01272-3

Dean, T. (2021, February 2). *Ethics explainer: Moral relativism.* The Ethics Centre. https://ethics.org.au/ethics-explainer-moral-relativism/

deGrasse Tyson, N. (2012). The case for space: why we should keep reaching for the stars. *Foreign Affairs, 91*(2), 22–33. http://www.jstor.org/stable/23217218

deGrasse Tyson, N. (2023). *What taxpayers gained going to the moon.* [Video]. YouTube. https://www.youtube.com/watch?v=Oe1NyLWGmVI

Deichmann, J., Ebel, E., Heineke, K., Heuss, R., Kellner, M., & Steiner, F. (2023, January 6). *Autonomous driving's future: Convenient and connected.* McKinsey & Company. https://www.mckinsey.com/industries/automotive-and-assembly/our-insights/autonomous-drivings-future-convenient-and-connected

Design Against Crime Research Centre. (2019). *Cell furniture: designs for safety, wellbeing, and sustainability.* https://files.cargocollective.com/c1114017/Cell-Furniture-Catalogue-_FINAL_compressed.pdf

Dick, P. K. (1956). The Minority Report. *Fantastic Universe Science Fiction.*

Dickerson, B. (2019, May 30). *The history of 20th-century (and earlier) wave pools.* WavePoolMag. https://wavepoolmag.com/the-complete-history-of-20th-century-wave-pools/#:~:text=In%20the%201870s%20Ludwig%20II,escape%20the%20pressures%20of%20kinghood

Dickerson, B. (2021, April 9). *Deep dive: so how much energy does a wave pool use?* WavePoolMag. https://wavepoolmag.com/deep-dive-so-how-much-energy-does-a-wave-pool-use/#:~:text=300%20to%20400%20kWh%20per,high%20in%20the%20same%20timeframe

Dignitas. (n.d.). *Who is Dignitas?* Dignitas. http://www.dignitas.ch/?lang=en

Dittmer, J. (1997). "Too busy to hate": Race, class, and politics in twentieth-century Atlanta. *The Georgia Historical Quarterly, 81*(1), 103–117.

Dotov, D. G., Nie, L., & Chemero, A. (2010). A demonstration of the transition from ready-to-hand to unready-to-hand. *PLoS ONE, 5*(3), 1–9. https://doi.org/10.1371/journal.pone.0009433

Doudna, J. & Sternberg, S. (2017). *A crack in creation: gene editing and the unthinkable power to control evolution.* Houghton Mifflin Harcourt.

Douglass, B. (1980). The Common Good and the Public Interest. *Political Theory, 8*(1), 103–117. http://www.jstor.org/stable/190769. https://doi.org/10.1177/009059178000800108

Driver, J. (2014, September 22). *The history of utilitarianism.* The Stanford Encyclopedia of Philosophy (Winter 2022 Edition). https://plato.stanford.edu/archives/win2022/entries/utilitarianism-history/

Dunham, W. (2022, July 19). Scientists uncover history of 'ridiculously charming' penguins. *Reuters.* https://www.reuters.com/lifestyle/science/scientists-uncover-history-ridiculously-charming-penguins-2022-07-19/#:~:text=The%20researchers%20said%20penguins%20evolved,becoming%20excellent%20swimmers%20and%20divers

Dunne, A. (1999). *Hertzian tales: electronic products, aesthetic experience, and critical design.* MIT Press.

Dunne, A. & Raby, F. (n.d.). *Critical design FAQ.* Dunne and Raby. https://dunneandraby.co.uk/content/bydandr/13/0

Engstrom, K. V., & van Ginneken, E. F. J. C. (2022). Ethical prison architecture: a systematic literature review of prison design features related to wellbeing. *Space and Culture,* 25(3), 479-503. https://doi.org/10.1177/12063312221104211

EPA. (n.d.). *Climate change and social vulnerability in the United States: a focus on six impacts. U.S. Environmental Protection Agency.* EPA. https://www.epa.gov/cira/social-vulnerability-report

European Parliament. (2023, April 27). *Ending fast fashion: tougher rules to fight excessive production and consumption.* European Parliament. https://www.europarl.europa.eu/news/en/press-room/20230424IPR82040/ending-fast-fashion-tougher-rules-to-fight-excessive-production-and-consumption

Everytown For Gun Safety. (2023, December 15). *U.S. Surpasses 2022 Total Unintentional Shootings by Children.* Everytown For Gun Safety. https://everytownsupportfund.org/press/u-s-surpasses-2022-total-unintentional-shootings-by-children/

Fair Trials. (2023, June 16). *EU Parliament approves landmark AI law.* Fair Trials. https://www.fairtrials.org/articles/news/eu-parliament-approves-landmark-ai-law/

Fekete, O.B., & Szilágyi, K.M. (2021) Contemporary zoo design and the historical treasure of garden art: Tiergarten Schönbrunn. *Journal of Landscape Architecture, 16,* 82-93, DOI: 10.1080/18626033.2021.2046809

Fenster, A. (2020, October 13). New data: solitary confinement increases risk of premature death after release. *Prison Policy Initiative.* https://www.prisonpolicy.org/blog/2020/10/13/solitary_mortality_risk/

Fiby, M. (2008). Trends in zoo design – changing needs in keeping wild animals for a visiting audience. *The International Review of Landscape Architecture and Urban Design.* https://zoolex.org/page/publications/fiby/trends_in_zoodesign

Fisher, T. (2019). *The architecture of ethics.* Routledge. https://doi.org/10.4324/9781351065740

Fisher, T. & Gamman, L. (2019). *Tricky design: the ethics of things.* Bloomsbury Visual Arts. https://doi.org/10.5040/9781474277211

Food Print. (2023, February 28). *How industrial agriculture affects our air.* Food Print. https://foodprint.org/issues/how-industrial-agriculture-affects-our-air/

Forcing functions. (n.d.). *Interaction Design Foundation.* https://www.interaction-design.org/literature/book/the-glossary-of-human-computer-interaction/forcing-functions

Forsythe, B. (1984). Prisons and panopticons. *Social Policy & Administration, 18*(1), 68-88. https://doi.org/10.1111/j.1467-9515.1984.tb00197.x

Foucault M. (1977). *Discipline and punish: the birth of the prison (1st American).* Pantheon Books.

Foust, J. (2019, April 16). *Pepsi drops plans to use orbital billboard.* Space News. https://spacenews.com/pepsi-drops-plans-to-use-orbital-billboard/

Fowler, G. (2022, March 31). *Am I being tracked? Anti-stalking tech from Apple, Tile falls short.* The Washington Post. https://www.washingtonpost.com/technology/2022/03/31/airtags-stalking/

Fraieli, A. (2021, May 10). *The cost to criminalize homelessness.* The Homeless Voice. https://homelessvoice.org/the-cost-to-criminalize-homelessness/

Franklin D. Roosevelt Presidential Library and Museum. (n.d.). *FDR and the Four Freedoms Speech.* https://www.fdrlibrary.org/four-freedoms

Fredrickson, D. & Siljander, R. (2002). *Racial profiling.* U.S. Department of Justice Office of Justice Programs. https://www.ojp.gov/ncjrs/virtual-library/abstracts/racial-profiling-0

Friedman, B., & Hendry, D. G. (2019). *Value sensitive design: shaping technology with moral imagination.* MIT Press. https://doi.org/10.7551/mitpress/7585.001.0001

Fry, H. (2018). *Hello world: being human in the age of algorithms.* W.W. Norton & Company.

Gaebler, J., Cai, W., & Goel, S. (2022, September 15). *Police stop Black drivers more often than Whites. We found out why.* The Washington Post. https://www.washingtonpost.com/politics/2022/09/15/driving-while-black-racial-discrimination-traffic-tickets/

Gallop, J. D. (2018, June 22). *No charges for 5 teens who mocked and filmed drowning man, Jamel Dunn, in a Cocoa pond.* Florida Today. https://www.floridatoday.com/story/news/2018/06/22/no-charges-year-after-teens-mocked-drowning-man-prosecutors-rule-out-filing-charges/723259002/

Galton, F. (1865). Hereditary talent and character. *Macmillan's Magazine, 12,* 157-166.

Gandhi, M. (2018). *An autobiography or the story of my experiments with truth: a critical edition.* New Haven: Yale University Press. https://doi.org/10.12987/9780300238402

Gender Shades Project. (n.d.). *Gender Shades.* http://gendershades.org/overview.html

Genoways, T. (2014). *The chain: farm, factory, and the fate of our food* (First edition). Harper, an imprint of Harper Collins Publishers.

Georgetown Law Center on Privacy & Technology. (2016, October 18). *The perpetual line-up.* https://www.perpetuallineup.org/findings/racial-bias

Gibson, J.J. (1979). *The ecological approach to visual perception.* Boston: Houghton Mifflin.

Gigante, M. (1993). Virtual reality: definitions, history, and applications. *Virtual Reality Systems,* 3-14. https://doi.org/10.1016/B978-0-12-227748-1.50009-3

Gogoll, J. et al. (2021). Ethics in the software development process: from codes of conduct to ethical deliberation. *Philos. Technol, 34,* 1085–1108. https://doi.org/10.1007/s13347-021-00451-w

Goldberg, N. (2023, May 8). *Column: high-speed, high adrenaline car chases are an L.A. staple. But are they worth the risks?* Los Angeles Times. https://www.latimes.com/opinion/story/2023-05-08/los-angeles-police-department-car-chase-police-pursuit-deaths-injuries

Goldstick, J., Cunningham, R., & Carter, P. (2022). Current causes of death in children and adolescents in the United States. *New England Journal of Medicine, 20,* 1955-1956. https://www.nejm.org/doi/full/10.1056/nejmc2201761

Goodman, P. (1969). "Can Technology Be Humane?" *New York Review of Books, 13*(9).

Governors Highway Safety Association. (2023, March). *Motorcyclists.* https://www.ghsa.org/state-laws/issues/Motorcyclists#:~:text=Illinois%2C%20Iowa%20and%20New%20Hampshire,not%20have%20motorcycle%20helmet%20laws

Grant, J. & Fox, F. (1992). Understanding the role of the designer in society. *Journal of Art & Design Education, 11,* 77-87. https://doi.org/10.1111/j.1476-8070.1992.tb00689.x

Gray, R. (2007, April 22). *How we made the Chernobyl rain.* The Telegraph. https://www.telegraph.co.uk/news/worldnews/1549366/How-we-made-the-Chernobyl-rain.html

Gregersen, L. (2022, August 11). *How a small Canadian city took on chronic homelessness.* Next City. https://nextcity.org/urbanist-news/medicine-hat-alberta-canada-city-chronic-homelessness

Gun Violence Archive. (n.d.). *GVA 10 year review.* https://www.gunviolencearchive.org/

Haberman, C. (2018, December 2). *For shame: a brief history of the perp walk.* The New York Times. https://www.nytimes.com/2018/12/02/us/perp-walk.html

Hall, L. (2019, July 15). *Going to the moon was hard – but the benefits were huge, for all of us.* NASA. https://www.nasa.gov/technology/tech-transfer-spinoffs/going-to-the-moon-was-hard-but-the-benefits-were-huge-for-all-of-us/

Hampton, L. (2023). Jeremy Bentham. In R.Y. Chappell, D. Meissner, and W. MacAskill (eds.), *An introduction to utilitarianism.* https://www.utilitarianism.net/utilitarian-thinker/jeremy-bentham

Harari, Y. N. (2024). *Nexus: a brief history of information networks from the Stone Age to AI. First U.S. edition.* New York, Random House.

Harvey, C. (2021, March 16). *Eight states are seeding clouds to overcome megadrought.* Scientific American. https://www.scientificamerican.com/article/eight-states-are-seeding-clouds-to-overcome-megadrought/

Heaven, W. (2020, July 17). *Predictive policing algorithms are racist. They need to be dismantled.* MIT Technology Review. https://www.technologyreview.com/2020/07/17/1005396/predictive-policing-algorithms-racist-dismantled-machine-learning-bias-criminal-justice/

Helliwell, J. F., Huang, H., Norton, M., Goff, L., & Wang, S. (2023). World happiness, trust, and social connections in times of crisis. In *World Happiness Report* 2023 (11th ed., Chapter 2). Sustainable Development Solutions Network. https://worldhappiness.report/ed/2023/world-happiness-trust-and-social-connections-in-times-of-crisis/#ranking-of-happiness-2020-2022

Herold, E. (2024). *Robots and the people who love them: holding on to our humanity in an age of social robots.* First edition. New York, St. Martin's Press.

Hersh, M. (2015). Ethical engineering: definitions, theories, and techniques. In M. Hersh (Ed.) *Ethical Engineering for International Development and Environmental Sustainability.* Springer-Verlag. https://doi.org/10.1007/978-1-4471-6618-4_2

Hickman, M. (2020, November 2). *Copenhill, BIG's skiable waste-to-energy power plant, gets sweeping new photos from Hufton + Crow.* The Architect's Newspaper. https://www.archpaper.com/2020/11/copenhill-bigs-skiable-waste-to-energy-power-plant-gets-sweeping-new-photos-from-hufton-crow/

Hippo Roller. (n.d.). *Revolutionizing water access: the resurgence of Hippo Roller technology.* Hippo Roller. https://hipporoller.org

Hoffman, H. et al., (2022). Virtual reality as an adjunctive non-pharmacologic analgesic for acute burn pain during medical procedures. *Annals of Behavioral Medicine, 4*(2), 183-191. https://doi.org/10.1007/s12160-010-9248-7

HOK. (n.d.). *The future of courthouse design.* HOK. https://www.hok.com/ideas/research/the-future-of-courthouse-design/

Holdorf, W.J. (1987, February 5). *Seat belt law an assault on freedom.* The Chicago Tribune. https://www.chicagotribune.com/news/ct-xpm-1987-02-05-8701090791-story.html

Holland, O. (2020, February 5). *PETA ends "I'd rather go naked" anti-fur campaign after three decades.* CNN. https://www.cnn.com/style/article/peta-naked-fur-campaign-ends/index.html

How does the Mosquito alarm work? (n.d.). *Mosquito loitering solutions.* https://mosquitoloiteringsolutions.com/faq/how-does-it-work/

How Mosquito helps in the community. (n.d.). *Mosquito loitering solutions.* https://mosquitoloiteringsolutions.com/how-mosquito-has-helped/in-the-community/

How to stop anti-social behavior. (n.d.). *Mosquito loitering solutions.* https://mosquitoloiteringsolutions.com/why-mosquito/anti-social-behaviour/

Howard, J. (1777). *The state of the prisons in England and Wales: with preliminary observations, and an account of some foreign prisons.* Printed by William Eyres, MDCCLXXVII.

Huling, T. (2002). Invisible punishment: the collateral consequences of mass imprisonment. *The New Press.* https://www.prisonpolicy.org/scans/building.html

Human Rights Watch. (2023, February 14). *US: New policy on autonomous weapons flawed.* Human Rights Watch. https://www.hrw.org/news/2023/02

/14/us-new-policy-autonomous-weapons-flawed?gad_source=1&gclid=CjwKCAiAopuvBhBCEiwAm8jaMR32yi1cMepw2JbYWekty8_ITP69BbC0PHbnJ-3H6i8Z7UmBam2aCBoC6zsQAvD_BwE

Imagine humane technology. (2021, August 17). *Center for Humane Technology.* https://www.humanetech.com/youth/imagine-humane-technology

IMDB. (n.d.). *Jeff Goldblum: Malcolm.* https://www.imdb.com/title/tt0107290/characters/nm0000156

Insurance Institute for Highway Safety. (2023, July). *Seat belt laws.* Insurance Institute for Highway Safety. https://www.iihs.org/topics/seat-belts#:~:text=With%20the%20exception%20of%20New,and%20the%20District%20of%20Columbia

Interaction 18. (n.d.). *The Oppenheimer Moment.* Interaction. https://interaction18.ixda.org/program/keynote--alan-cooper/

Invasive Species Advisory Committee. (2016, December 6). *Invasive species impacts on infrastructure.* https://www.doi.gov/sites/doi.gov/files/uploads/isac_infrastructure_white_paper.pdf

J. Robert Oppenheimer, atomic bomb pioneer, dies. (1967, February 19). *The New York Times.* https://archive.nytimes.com/www.nytimes.com/learning/general/onthisday/bday/0422.html

Jacobs, N., & Huldtgren, A. (2021). Why value sensitive design needs ethical commitments. *Ethics & Information Technology, 23*(1), 23–26. https://doi.org/10.1007/s10676-018-9467-3

Jones, R.P. & Cox, D. (2011). *Few Americans see natural disasters a sign from god.* PRRI. http://www.prri.org/research/few-americans-see-earthquakes-floods-and-other-natural-disasters-a-sign-from-god-2/

Jordan, D. (2023, September 8). *Elon Musk says he withheld Starlink over Crimea to avoid escalation.* BBC News. https://www.bbc.com/news/world-europe-66752264

Kamiya, G. (2021, February 2). *The holier-than-thou crusade in San Francisco.* The Atlantic. https://www.theatlantic.com/ideas/archive/2021/02/san-francisco-renaming-spree/617894/

Kampourakis, K. (2020). Students' "teleological misconceptions" in evolution education: why the underlying design stance, not teleology per se, is the problem. *Evo Edu Outreach, 13*(1). https://doi.org/10.1186/s12052-019-0116-z

Kamprani, K. (n.d.). *The Uncomfortable.* https://www.theuncomfortable.com/about/

Kang-Brown, J. & Subramanian, R. (2017, June). *Out of sight: the growth of jails in rural America.* Vera Institute of Justice. https://www.vera.org/downloads/publications/out-of-sight-growth-of-jails-rural-america.pdf

Keim, B. (2018, August 22). *Is New Zealand's war on invasive predators an expensive distraction?* Anthropocene. https://www.anthropocenemagazine.org/2018/08/new-zealand-predator-free-critique/?gclid=CjwKCAiA-vOsBhAAEiwAIWR0TQjDGM5UcWUN6Dh3FmechhPgGrMMbJsKt0bcMIOPteimD-W931LQeBoCMGcQAvD_BwE

Kennedy, J. (1962, September). *We choose to go to the moon.* Rice University. https://www.rice.edu/jfk-speech

Kessler, G. (2021, November 10). *Robert Moses and the sage of the racist parkway bridges.* The Washington Post. https://www.washingtonpost.com/politics/2021/11/10/robert-moses-saga-racist-parkway-bridges/

Khurshudyan, I., Ilushina, M., & Khudov, K. (2022, December 2). *Russia and Ukraine are fighting the first full-scale drone war.* The Washington Post. https://www.washingtonpost.com/world/2022/12/02/drones-russia-ukraine-air-war/

Klosterman, C. (2014, June 20). *So you think you can steal?* The New York Times Magazine. https://www.nytimes.com/2014/06/22/magazine/so-you-think-you-can-steal.html

Koonin, S. (2021). *Unsettled: What climate science tells us, what it doesn't, and why it matters.* BenBella Books, Inc.: Dallas.

Lanier, J. (2017). *Dawn of the new everything: encounters with reality and virtual reality.* Henry Holt and Company.

Lantigua-Williams, J. (2016, December 5). *The link between race and solitary confinement.* The Atlantic. https://www.theatlantic.com/politics/archive/2016/12/race-solitary-confinement/509456/

LAPD. (2023, March 14). *Officer involved shooting occurs during bait car operation.* LAPD. https://www.lapdonline.org/newsroom/officer-involved-shooting-occurs-during-bait-car-operation/

Laranjeira, C., Dixe, M. A., Querido, A., & Stritch, J. M. (2022). Death cafés as a strategy to foster compassionate communities: Contributions for death and grief literacy. *Frontiers in Psychology, 13,* 986031. https://doi.org/10.3389/fpsyg.2022.986031

Larrea, M. (2023, February 13). *What is a gated community? Pros & cons of living in gated communities.* Palm Paradise Realty Group. https://www.palmparadiserealestate.com/blog/gated-community-pros-cons/

Latour, B. (2000). *Pandora's hope: essays on the reality of science studies.* Harvard University Press.

Lee, H. N. (1928). Morals, Morality, and Ethics: Suggested Terminology. *International Journal of Ethics, 38*(4), 450–466. http://www.jstor.org/stable/2377964. https://doi.org/10.1086/207693

Levin, N. (2019). "Introduction to Philosophy and the Ship of Theseus." In *Ancient Philosophy Reader, An Open Educational Resource.* NGE Far Press. https://human.libretexts.org/Bookshelves/Philosophy/Book%3A_Ancient_Philosophy_Reader_(Levin)/01%3A_The_Start_of_Western_Philosophy_and_the_Pre-Socratics/1.01%3A_Introduction_to_Philosophy_and_the_Ship_of_Theseus

Liberty Human Rights. (2019, February 4). *Liberty report exposes police forces' use of discriminatory data to predict crime.* https://www.libertyhumanrights.org.uk/issue/liberty-report-exposes-police-forces-use-of-discriminatory-data-to-predict-crime/

Lidge, E.F. (2006). Perp walks and prosecutorial ethics. *Nevada Law Review,* 7(26). https://scholars.law.unlv.edu/cgi/viewcontent.cgi?article=1419&context=nlj

Liepert, A.E., Berndtson, A.E., & Hill, L.L. (2022). Association of 30-ft US-Mexico border wall in San Diego with increased migrant deaths, trauma center admissions, and injury severity. *JAMA Surg, 157*(7), 633–635. https://doi.org/10.1001/jamasurg.2022.1885

Lillehammer, H. (2023). Keeping track of your trolleys: origins and destinations. In H. Lillehammer (Ed.), *The trolley problem* (Classic Philosophical Arguments, pp. 7-24). Cambridge: Cambridge University Press. https://doi.org/10.1017/9781009255615.002

Lloyd, S. & Sreedhar, S. (2022, September 12). Hobbes's moral and political philosophy. *The Stanford Encyclopedia of Philosophy (Winter 2022 Edition).* https://plato.stanford.edu/entries/hobbes-moral/#StaNat

Lo, A. (2017, December 21). *The debate: Is hostile architecture designing people – and nature – out of cities?* CNN. https://www.cnn.com/style/article/new-dean-harvey-james-furzer-hostile-architecture-debate/index.html

Lockton, D. (n.d.). *Architectures of control in the built environment.* Architectures. https://architectures.danlockton.co.uk/architectures-of-control-in-the-built-environment/

Luckey, P. (2022, November 6). *If you die in the game, you die in real life.* The Blog of Palmer Luckey. https://palmerluckey.com/if-you-die-in-the-game-you-die-in-real-life/

Lung, N. (2023, October 20). *Most New York Airbnb applications don't meet city requirements.* Bloomberg. https://www.bloomberg.com/news/articles/2023-10-20/most-new-york-airbnb-applications-don-t-meet-city-requirements

Madden, K. Kent, F., & Peinhardt, K. (2021, November 8). *How seating shapes welcoming cities.* Social Life Project. https://www.sociallifeproject.org/have-a-seat-how-seating-shapes-welcoming-cities/

Manaugh, G. (2016). *A burglar's guide to the city* (First edition). Farrar, Straus and Giroux.

Mandela, N. (2009). *Long walk to freedom.* New York: Flash Point/Roaring Brook Press.

Markkula Center for Applied Ethics. (2018, June 22). *An ethical toolkit for engineering/design practice.* https://www.scu.edu/ethics-in-technology-practice/ethical-toolkit/

Markkula Center for Applied Ethics. (2021, November 8). *A framework for ethical decision making.* https://www.scu.edu/ethics/ethics-resources/a-framework-for-ethical-decision-making/

Martindale, D. (2023, April 26). *Bringing back woolly mammoths and dodos is a bad idea.* Vox. https://www.vox.com/future-perfect/23696294/de-extinction-colossal-biosciences-woolly-mammoth-dodo-ethics

Mason, R. O. (1986). Four Ethical Issues of the Information Age. *MIS Quarterly, 10*(1), 5–12. https://doi.org/10.2307/248873

McDowell, J. (2024, April 14). *GCAT: general catalog of artificial space objects.* https://planet4589.org/space/gcat/index.html

Merchant, B. (2023). *Blood in the machine: the origins of the rebellion against big tech.* Little, Brown, and Company.

Merchant, N. (n.d.). *IoT technologies explained: History, examples, risks, & future.* Vision of Humanity. https://www.visionofhumanity.org/what-is-the-internet-of-things/#:~:text=The%20term%20'Internet%20of%20Things,them%20through%20a%20supply%20chain

Metzl, J. F. (2019). *Hacking Darwin: genetic engineering and the future of humanity.* Sourcebooks, Inc.

Mitchell, WJT (1998). The violence of public art. In Ostrow, S. (Ed.), *Beauty is nowhere: ethical issues in art and design.* Routledge. https://doi.org/10.4324/9780203871423

Monteiro, M. (2019). *Ruined by design: How designers destroyed the world, and what we can do to fix it.* Independently published.

Moore, A. (2022, September 5). *'I didn't want it anywhere near me': how the Apple AirTag became a gift to stalkers.* The Guardian. https://www.theguardian.com/technology/2022/sep/05/i-didnt-want-it-anywhere-near-me-how-the-apple-airtag-became-a-gift-to-stalkers

Moore, T. (2019, July17). *Inside the design evolution of the modern zoo.* Bloomberg. https://www.bloomberg.com/news/articles/2019-07-17/the-history-of-humans-looking-at-animals-in-zoos

Moreno, C. (2021, June 21). *Decades after the Tulsa race massacre, urban 'renewal' sparked Black Wall Street's second destruction.* Smithsonian Magazine. https://www.smithsonianmag.com/history/black-wall-streets-second-destruction-180977871/

Morgan, R.E. & Truman, J.L. (2022, February). *Stalking victimization, 2019.* U.S. Department of Justice. https://bjs.ojp.gov/content/pub/pdf/sv19.pdf

Murphy, D. et al. (2023). Metals from spacecraft reentry in stratospheric aerosol particles. *PNAS, 120*(43). https://www.pnas.org/doi/full/10.1073/pnas.2313374120. https://doi.org/10.1073/pnas.2313374120

Murphy, T. (2015, July 22). *Ben Carson says prison is so comfy some people never want to leave.* Mother Jones. https://www.motherjones.com/politics/2015/07/ben-carson-says-prisons-are-too-comfy/. https://doi.org/10.7748/nm.22.8.19.s19

Murray, C. (2019, March 28). *It's time for architects to choose ethics over aesthetics.* dezeen. https://www.dezeen.com/2019/03/28/opinion-christine-murray-climate-change/

Murrow, E. (1955, January 4). *A Conversation with J. Robert Oppenheimer.* See It Now. https://famous-trials.com/oppenheimer/2694-edward-r-murrow-interviews-oppenheimer-see-it-now

Narcotta, J. & Ablondi, B. (2022, June 22). *Strategy analytics: Amazon's Ring remained atop the video doorbell market in 2021.* Business Wire. https://www.businesswire.com/news/home/20220622005023/en/Strategy-Analytics-

Amazons-Ring-Remained-atop-the-Video-Doorbell-Market-in-2021#:~:text =These%20five%20companies%20combined%20to,to%20consumers%20glo bally%20in%202021

Nelson, G. (1977). *How to see: visual adventures in a world God never made (1st ed).* Little Brown.

Newcomb, T. (2023, January 30). *Scientists are reincarnating the woolly mammoth to return in 4 years.* Popular Mechanics. https://www.popularmechanics.com/science/animals/a42708517/scientists-reincarnating-woolly-mammoth/

Nielsen, K. (2004). *The radical lives of Helen Keller.* New York University Press.

Nikke Asia. (2023, July 7). *Japan police to test AI-equipped cameras in protecting VIPs.* https://asia.nikkei.com/Politics/Japan-police-to-test-AI-equipped-cameras-in-protecting-VIPs

Nilson, E. et al. (2022). Premature deaths attributable to the consumption of ultraprocessed foods in Brazil. *Global Health Promotion and Prevention, 64*(1), 129-136. https://doi.org/10.1016/j.amepre.2022.08.013

Nolen, S. (2002, November 5). *Designing zoo habitats that promote animal well-being.* American Veterinary Medical Association. https://www.avma.org/javma-news/2002-12-01/designing-zoo-habitats-promote-animal-well-being #:~:text=Research%20has%20also%20shown%20that,for%20and%20catch% 20their%20food

Norman, D. A. (1988). *The psychology of everyday things.* Basic Books.

Novak B. J. (2018). De-Extinction. *Genes, 9*(11), 548. https://doi.org/10.3390/ge nes9110548

NYCLU. Stop-and-frisk data. *ACLU of New York.* https://www.nyclu.org/en/stop-and-frisk-data

O'Brien, M. (2022, September 2). *Years late and billions over budget, NASA's most powerful rocket finally set for takeoff.* PBS. https://www.pbs.org/newshour/show /years-late-and-billions-over-budget-nasas-most-powerful-rocket-finally-set-for-takeoff

O'Connell, M. (2017). *To be a machine: adventures among cyborgs, utopians, hackers, and the futurists solving the modest problem of death.* Doubleday.

Odenbaugh, J. (2023). Philosophy and ethics of de-extinction. *Cambridge Prisms: Extinction, 1*(7). doi:10.1017/ext.2023.4

Olmos, S. (2023, April 23). *LA is blasting classical music to deter homeless people from gathering at metro stations.* NPR. https://www.npr.org/2023/04/23/117 1494452/la-is-blasting-classical-music-to-deter-homeless-people-from-gathering-at-metro-

Ong, P. (n.d.). *Priscilla Ong studio.* https://www.priscillaong.com/

Overkamp, L. (2019, May 30). *Daily ethical design.* A List Apart. https://alistapart.com/article/daily-ethical-design/

Overmyer-Velazquez, M. (2011). *Beyond La Frontera: the history of Mexico-U.S. migration.* Oxford University Press.

Palm, E. & Hansson, S.O. (2006). The case for ethical technology assessment (eTA). *Technological Forecasting and Social Change, 73*(5), 543-558. https://doi.org/10.1016/j.techfore.2005.06.002

Pandya, D. & Kalra, A. (2023, July 20). *India's Adani to convert Mumbai slum into modern city despite challenges.* Reuters. https://www.reuters.com/world/india/indias-adani-convert-mumbai-slum-into-modern-city-despite-challenges-2023-07-20/

Papanek, V. (1998). Toward the spiritual in design. In Ostrow, S. (Ed.), *Beauty is nowhere: ethical issues in art and design.* Routledge. https://doi.org/10.4324/9780203871423

Parshina-Kottas, Y., Singhvi, A., Burcj, A., Griggs, T., Grondahl, M., Huang, L., Wallace, T., White, J., & Williams, J. (2021, May 24). *What the Tulsa race massacre destroyed.* The New York Times. https://www.nytimes.com/interactive/2021/05/24/us/tulsa-race-massacre.html

Perez, S. (2021, September 16). *Tile secures $40 million to take on Apple AirTag with new products.* Tech Crunch. https://techcrunch.com/2021/09/16/tile-secures-another-40-million-to-take-on-apple-airtag-with-new-products/

PETA. (n.d.). *Don't zoos help to preserve endangered species?* PETA. https://www.peta.org/about-peta/faq/dont-zoos-help-to-preserve-endangered-species/

PETA. (n.d.). *How to tell if a place is a REAL animal sanctuary.* PETA. https://www.peta.org/features/real-animal-sanctuary-zoo/

PETA. (n.d.). *Our mission statement.* PETA. https://www.peta.org/about-peta/

PETA. (n.d.). *Why animal rights?* PETA. https://www.peta.org/about-peta/why-peta/why-animal-rights/

Pew Research Center. (2022, March 15). *Public expresses mixed views of U.S. response to Russia's invasion of Ukraine.* Pew Research. https://www.pewresearch.org/politics/2022/03/15/public-expresses-mixed-views-of-u-s-response-to-russias-invasion-of-ukraine/

Phelps, W. (2021). *On killing remotely: the psychology of killing with drones.* Little, Brown, and Company.

Plump, W. (n.d.). *Peter Singer on ethical politics.* Princeton Magazine. https://www.princetonmagazine.com/peter-singer-on-ethical-politics/#:~:text=As%20both%20a%20surfer%20and,waves%20are%20good%20or%20not

Poinar, G. O., Hess, R. (1982). Ultrastructure of 40-Million-Year-Old Insect Tissue. *Science, 215.* 1241-1242. DOI:10.1126/science.215.4537.1241

Polakova, S. & Cruchade, A. (2018, October 24). *Can speculative design make UX better?* Medium. https://uxplanet.org/can-speculative-design-make-ux-better-design-trend-4-4-ce8d13148e5d

Pontin, M.W. (2008, March 25). *Weather engineering in China.* MIT Technology Review. https://www.technologyreview.com/2008/03/25/270084/weather-engineering-in-china/

Portes, J. (2019). The Economics of Migration. *Contexts, 18*(2), 12-17. https://doi.org/10.1177/1536504219854712

Postman, N. (1993). *Technopoly: the surrender of culture to technology.* 1st Vintage Books ed. New York, Vintage Books.

Postman, N. (1999). *Building a bridge to the 18th century: how the past can improve our future.* Alfred A. Knopf: Distributed by Random House.

Prasow, A. (2020, December 14). *US architects join movement against capital punishment, torture.* Human Rights Watch. https://www.hrw.org/news/2020/12/14/us-architects-join-movement-against-capital-punishment-torture#:~:text=On%20December%2011%2C%20the%20Board,of%20prisoners%20for%2022%20hours

Pratten, J.D., Wright, D.L. (2013). Unethical products. In: Idowu, S.O., Capaldi, N., Zu, L., Gupta, A.D. (eds) *Encyclopedia of Corporate Social Responsibility.* Springer, Berlin, Heidelberg. https://doi.org/10.1007/978-3-642-28036-8_598

Project for Public Spaces. (2008, December 31). *Waste receptacles.* PPS. https://www.pps.org/article/wastereceptacles

Rascoe, A. & Kocherga, A. (2023, October 8). *The Biden administration is building a controversial part of the border wall in Texas.* NPR News. https://www.npr.org/2023/10/08/1204545880/the-biden-administration-is-building-a-controversial-part-of-the-border-wall-in-

Rainie, L., Funk, C., Anderson, M., & Tyson, A. (2022, March 17). *Americans are closely divided over editing a baby's genes to reduce serious health risk.* Pew Research Center. https://www.pewresearch.org/internet/2022/03/17/americans-are-closely-divided-over-editing-a-babys-genes-to-reduce-serious-health-risk/

Rainie, L., Funk, C., Anderson, M., & Tyson, A. (2022, March 17). *Public more likely to see facial recognition use be police good, rather than bad for society.* Pew Research Center. https://www.pewresearch.org/internet/2022/03/17/public-more-likely-to-see-facial-recognition-use-by-police-as-good-rather-than-bad-for-society/

Ramirez, E. J. (2022). *The ethics of virtual and augmented reality: Building worlds.* Routledge. https://doi.org/10.4324/9781003042228

RAWTools. (n.d.). *Forging peace. Disarming hearts.* RAW Tools. https://rawtools.org/

Recycling Today. (2019, November). *The woes of poorly sorted waste – an industry perspective.* Recycling Today. https://www.recyclingtoday.com/article/the-woes-of-poorly-sorted-waste--an-industry-perspective/

Rees, P. A. (2023). *Zoo studies: living collections, their animals, and visitors.* Cambridge: Cambridge University Press. https://doi.org/10.1017/9781108566049

Reich, R. B. (2018). *The common good. First edition.* New York, Alfred A. Knopf.

René Magritte: Belgian, 1898-1967. (n.d.). Museum of Modern Art. https://www.moma.org/artists/3692#fn:1

Reuters. (2023, September 12). *US-Mexico border is world's deadliest land migration route, IOM finds.* Reuters. https://www.reuters.com/world/us-mexico-border-is-worlds-deadliest-land-migration-route-iom-finds-2023-09-12/

Reyes, P. (n.d.). *Palas por pistolas.* Pedro Reyes. http://pedroreyes.net/08.php?szLang=en&Area=work&SubArea=08

Richards, Z. (2023, December 19). *Trump doubles down on immigrant 'blood' remark, says he 'never read Mein Kampf.'* NBC News. https://www.nbcnews.com/politics/donald-trump/trump-doubles-immigrant-blood-remark-says-never-read-mein-kampf-rcna130535

Rivera, N. (2020). *Painted over in brown: border art as visual discourse of resistance.* UNM Digital Repository. https://digitalrepository.unm.edu/cgi/viewcontent.cgi?article=1140&context=cj_etds

Roach, M. (2021). *Fuzz: when nature breaks the law.* New York, NY, W.W. Norton & Company.

Rodgers, B. (2023). *Cloak of charity: studies in eighteenth-century philanthropy.* Taylor & Francis Group. https://doi.org/10.4324/9781003459392

Roe, K., McConney, A., & Mansfield, C. F. (2014). How do zoos 'talk' to their general visitors? Do visitors 'listen'? A mixed method investigation of the communication between modern zoos and their general visitors. *Australian Journal of Environmental Education, 30*(2), 167–186. https://www.jstor.org/stable/26422865. https://doi.org/10.1017/aee.2015.1

Rose, C. (2015, March 22). *Astrophysicist Neil deGrasse Tyson's one-man mission.* CBS News. https://www.cbsnews.com/news/neil-degrasse-tyson-astrophysicist-charlie-rose-60-minutes/

Rosenberg, M., Cooke, K., & Trotta, D. (2022, July 25). *The border's toll.* Reuters. https://www.reuters.com/investigates/special-report/usa-immigration-border-deaths/

Rosenberger, R. (2017). *Callous Objects: Designs Against the Homeless.* Minneapolis, MN: University of Minnesota Press. https://doi.org/10.5749/9781452958538

Ross, S. (n.d.). *2005-2006 Archisuits.* Insecure Spaces. https://insecurespaces.net/archisuits-2005-2006/

Rousseau J.-J. & Cress D. A. (1987). *Basic political writings.* Hackett Publishing Company.

Rybczynski, W. (2013). *How architecture works: A humanist's toolkit.* Farrar, Strous, and Giroux. New York.

Sabin Center for Climate Change Law. (2024). *About.* Climate Case Chart. http://climatecasechart.com/about/

Sagan, C. (1985). *We are all in this greenhouse together.* [Video]. YouTube. https://www.youtube.com/watch?v=EuR9CzWcCfc

Salvador Dali: The persistence of memory. (n.d.). Museum of Modern Art. https://www.moma.org/collection/works/79018

Schön, D. A. (1983). *The reflective practitioner: how professionals think in action.* New York: Basic Books.

Science. (1997, May 6). *Watson urges "put Hitler behind us."* Science. https://www.science.org/content/article/watson-urges-put-hitler-behind-us

ScienceDaily. (2023, April 7). *Internet access must become human right or we risk ever-widening inequality.* ScienceDaily. https://www.sciencedaily.com/releases/2023/04/230407094243.htm

Seife, C. (2014). *Virtual unreality.* Penguin Group.

Selinger, E. (2012, July 23). *The philosophy of the technology of the gun.* The Atlantic. https://www.theatlantic.com/technology/archive/2012/07/the-philosophy-of-the-technology-of-the-gun/260220/

Shimron, Y. (2023, October 6). *U.S. faith groups do not view climate change as a crisis, new poll finds.* The Washington Post. https://www.washingtonpost.com/religion/2023/10/06/religion-climate-change-survey/

Similarweb. (n.d.). *Top websites ranking.* Similar web. https://www.similarweb.com/top-websites/lifestyle/fashion-and-apparel/

Sinclair, W. (1978, December 23). *U.S. and Mexico embroiled in dispute over 'tortilla curtain.'* The Washington Post. https://www.washingtonpost.com/archive/politics/1978/12/24/us-and-mexico-embroiled-in-dispute-over-tortilla-curtain/0f9b055a-5989-4600-8ecb-a642b8a14777/

Sinclair, U. (1906). *The Jungle.* New York, New American Library.

Singer, P. (2006, January 4). *Godless morality.* Project Syndicate. https://www.project-syndicate.org/commentary/godless-morality

Singer, P. (2009, May 18). *Isaac Asimov's laws of robotics are wrong.* Brookings. https://www.brookings.edu/articles/isaac-asimovs-laws-of-robotics-are-wrong/

Singer, P. (2009). *Animal liberation; The definitive classic of the animal movement.* Open Road Integrated Media, Inc.

Singer, P. (2011, July 13). *Moral progress and animal welfare.* Project Syndicate. https://www.project-syndicate.org/commentary/moral-progress-and-animal-welfare#:~:text=PRINCETON%20%E2%80%93%20Mahatma%20Gandhi%20acutely%20observed,mark%20of%20a%20civilized%20society.

Singer, P. (2016). *Ethics in the Real World: 82 Brief Essays on Things That Matter (REV-Revised).* Princeton University Press. https://doi.org/10.2307/j.ctvc77g39

Slater, M. (2020). The ethics of realism in virtual and augmented reality. *Frontiers in Virtual Reality, 1*(1). doi: 10.3389/frvir.2020.00001

Slum clearance. (1933, December 19). *The New York Times.* https://www.nytimes.com/1933/12/19/archives/slum-clearance.html?url=http%3A%2F%2Ftimesmachine.content-tagging.us-east-1-01.prd.dvsp.nyt.net%2Ftimesmachine%2F1933%2F12%2F19%2F105832327.html

Smart Security Pros. (2021, March 19). *Keep schools safe with advanced video surveillance.* Smart Security Pros. https://smartsecuritypros.com/keep-schools-safe-with-video-surveillance/

Soccer Stadium Digest. (n.d.). *2022 MLS attendance.* Soccer Stadium Digest. https://soccerstadiumdigest.com/2022-mls-attendance/

SolitaryWatch. (2023, May). *Calculating torture.* SolitaryWatch. https://solitarywatch.org/calculating-torture/

Southern Poverty Law Center. (n.d.). *Westboro Baptist Church.* https://www.splcenter.org/fighting-hate/extremist-files/group/westboro-baptist-church

Spagat, E. (2023, December 19). *Illegal crossings surge in remote areas as Congress and the White House weigh major asylum limits.* AP News. https://apnews.com/article/arizona-border-wall-breaches-asylum-immigration-beddd2ca0ffd02884ce0d0ad1785f538

Speculative EDU. (2019, July 23). *Critical about critical and speculative design.* https://speculativeedu.eu/critical-about-critical-and-speculative-design/

Speri, A. (2021, June 10). *The NYPD is still stopping and frisking black people at disproportionate rates.* The Intercept. https://theintercept.com/2021/06/10/stop-and-frisk-new-york-police-racial-disparity/

Stanford Center on China's Economy and Institutions. (2022, September 15). *Who supports social credit systems in China?* https://sccei.fsi.stanford.edu/china-briefs/who-supports-social-credit-systems-china

Stein, S.M. & Harper, T.L. (2013). Designing for meaning: the ethical designer's responsibility. In C. Basta & S. Moroni (Eds.), *Ethics, design and planning of the built environment* (pps. 143-166). Springer Dordrecht. https://doi.org/10.1007/978-94-007-5246-7_10

Structural racism and discrimination. (2023, February 23). *National Institute of Minority Health and Health Disparities.* https://www.nimhd.nih.gov/resources/understanding-health-disparities/srd.html

Sullivan, A. (2021, May 25). *U.S. freeways flattened Black neighborhoods nationwide.* Reuters. https://www.reuters.com/world/us/us-freeways-flattened-black-neighborhoods-nationwide-2021-05-25/

Surfrider Foundation. (n.d.). *Wave pools: environmental concerns outweigh their value.* Surfrider. https://www.surfrider.eu/learn/news/wave-pools-environmental-concerns-outweigh-their-value/

Swan, R. (2024, January 6). *The Golden Gate Bridge suicide nets are nearly finished. The milestone comes with warnings from the future.* San Francisco Chronicle. https://www.sfchronicle.com/sf/article/golden-gate-bridge-suicide-18589579.php

Szondy, M. & Fazekas, P. (2024). Attachment to robots and therapeutic efficiency in mental health. *Frontiers in Psychology, 15.* https://www.frontiersin.org/journals/psychology/articles/10.3389/fpsyg.2024.1347177/full. https://doi.org/10.3389/fpsyg.2024.1347177

Temperton, J. (2023, July 21). *'Now I am become death, the destroyer of worlds.' The story of Oppenheimer's infamous quote.* Wired. https://www.wired.com/story/manhattan-project-robert-oppenheimer/

Testa, J. (2022, September 1). *The people's republic of Shein.* The New York Times. https://www.nytimes.com/2022/09/01/style/shein-clothing.html

The Associated Press. (2022, August 21). *China plans to use chemicals to generate rain to protect its grain harvest.* NPR. https://www.npr.org/2022/08/21/1118683699/china-grain-harvest-cloud-seeding

The Atlanta Constitution. (1962, December 19). *Atlanta wall 03* [Photograph]. https://dilemmaxdotnet.files.wordpress.com/2021/02/1962-december-19-atlanta-wall-03.jpg

The Death Designer. (n.d.). *Designing a better deathcare for everyone.* The Death Designer. https://www.thedeathdesigner.com/

The Humane League. (2020, December 3). *Everything you should know about battery cages.* https://thehumaneleague.org/article/battery-cages

The Intercept. (n.d.). *Border Patrol Strategic Plan 1994 and Beyond.* https://www.documentcloud.org/documents/5987025-Border-Patrol-Strategic-Plan-1994-and-Beyond.html

The International Crime Prevention Through Environmental Design Association. (n.d.). *What is CPTED?* https://www.cpted.net/

The Nature Conservancy. (2019, October 14). *Python patrol: stopping a Burmese python invasion.* Nature. https://www.nature.org/en-us/about-us/where-we-work/united-states/florida/stories-in-florida/stopping-a-burmese-python-invasion/?gad_source=1&gclid=CjwKCAiA-vOsBhAAEiwAIWR0TYgB3jGDGjnqUgErsxm5N-venWNH-hEP5ZLeDM9N_8PDIOPpTWSggRoCUUsQAvD_BwE&gclsrc=aw.ds

The University of Texas at Austin Energy Institute. *The timeline and events of the February 2021 Texas electric grid blackouts.* https://energy.utexas.edu/research/ercot-blackout-2021

The White House. (2023, October 5). *Remarks by President Biden before national security team briefing on Ukraine.* https://www.whitehouse.gov/briefing-room/speeches-remarks/2023/10/05/remarks-by-president-biden-before-national-security-team-briefing-on-ukraine/

Thierer, A. (2024, October). *Defining "Technology."* Medium. https://medium.com/@AdamThierer/defining-technology-0c5c95c8360d

Thomas, S. (2011). Situated empathy: constructed theoretical discourse addressing the empathetic motivations shared by fashion design for sustainability, and the potential of socially engaged Buddhist ethics to inform design practice. Ph.D. Royal Melbourne Institute of Technology University. https://s3-ap-southeast-1.amazonaws.com/ap-st01.ext.exlibrisgroup.com/61RMIT_INST/storage/alma/78/9C/66/BF/AE/B1/8B/46/60/D5/1B/FC/A2/E1/7B/9E/Thomas.pdf?response-content-type=application%2Fpdf&X-Amz-Algorithm=AWS4-HMAC-SHA256&X-Amz-Date=20240119T185140Z&X-Amz-SignedHeaders=host&X-Amz-Expires=119&X-Amz-Credential=AKIAJN6NPMNGJALPPWAQ%2F20240119%2Fap-southeast-1%2Fs3%2Faws4_request&X-Amz-Signature=436f24d4410537aad5abfbcaf71cbea25c3a626a408dbff9a790b58345763cea

Thomas, S. (2018). *Fashion ethics.* Routledge. https://doi.org/10.4324/9781315109367

Thoreau, H. D. (1917). *Walden, or, Life in the woods.* Chicago, New York: Scott, Foresman and Company.

Thunberg, G. (2019). *Full speech to world leaders at UN Climate Action Summit.* [Video]. YouTube. https://www.youtube.com/watch?v=KAJsdgTPJpU

Tobón, S. (2022). Do better prisons reduce recidivism? Evidence from a prison construction program. *The Review of Economics and Statistics, 104*(6), 1256–1272. doi: https://doi-org.proxy1.library.jhu.edu/10.1162/rest_a_01007

Torrella, K. (2023, December 12). *We raise 18 billion animals a year to die – and then we don't even eat them.* Vox. https://www.vox.com/future-perfect/22890292/food-waste-meat-dairy-eggs-milk-animal-welfare

Tseng, P. E., & Wang, Y. H. (2021). Deontological or utilitarian? An eternal ethical dilemma in Outbreak. *International Journal of Environmental Research and Public Health, 18*(16), 8565. https://doi.org/10.3390/ijerph18168565

Tully, J. (1986, December 19). *Einstein letter sold to Forbes.* The Washington Post. https://www.washingtonpost.com/archive/lifestyle/1986/12/20/einstein-letter-sold-to-forbes/1f5d23ff-47e3-43ba-a7c0-0199643f1933/

Tyson, A. & Kennedy, B. (2023, October 25). *How Americans view future harms from climate change in their community and around the U.S.* Pew Research Center. https://www.pewresearch.org/science/2023/10/25/how-americans-view-future-harms-from-climate-change-in-their-community-and-around-the-u-s/

U.S. Department of Transportation. (2023, September 23). *Federal Aviation Administration.* https://www.faa.gov/sites/faa.gov/files/Report_to_Congress_Reentry_Disposal_of_Satellites.pdf

U.S. General Services Administration. (n.d.). *Alton Lennon Federal Building and U.S. Courthouse, Wilmington, NC.* U.S. General Services Administration. https://www.gsa.gov/real-estate/historic-preservation/explore-historic-buildings/find-a-building/all-historic-buildings/alton-lennon-federal-building-and-us-courthouse#:~:text=At%20the%20turn%20of%20the,entering%20or%20leaving%20the%20port.

Ufer, D. (2023, April 24). *Farm animal welfare policies cover breeding sows, veal calves, or laying hens in 14 U.S. states.* USDA Economic Research Service. https://www.ers.usda.gov/amber-waves/2023/april/farm-animal-welfare-policies-cover-breeding-sows-veal-calves-or-laying-hens-in-14-u-s-states/

UN Environment Programme. (2022, September 30). *Why you should rethink your next fashion purchase.* https://www.unep.org/news-and-stories/story/why-you-should-rethink-your-next-fashion-purchase

United Nations. (2011, October 18). *Solitary confinement should be banned in most cases, UN expert says.* United Nations. https://news.un.org/en/story/2011/10/392012

United Nations. (2021, October 3). *UN chief promotes 'enormous' benefits of greener cities.* United Nations. https://news.un.org/en/story/2021/10/1101992

United Nations. (n.d.). *Sustainability.* United Nations. https://www.un.org/en/academic-impact/sustainability

United States Environmental Protection Agency. (2024, April 10). *EPA history.* https://www.epa.gov/history

United States Geological Survey. (n.d.). *How many homes can an average wind turbine power?* https://www.usgs.gov/faqs/how-many-homes-can-average-wind-turbine-power

University of Michigan. (2019, June 20). *Moral concerns override desire to profit from finding a lost wallet.* ScienceDaily. www.sciencedaily.com/releases/2019/06/190620153447.htm

van de Poel, I. (2013). Translating Values into Design Requirements. In Michelfelder, D., McCarthy, N., Goldberg, D. (eds) *Philosophy and Engineering: Reflections on Practice, Principles and Process. Philosophy of Engineering and Technology,* vol 15. Springer, Dordrecht. https://doi-org.proxy1.library.jhu.edu/10.1007/978-94-007-7762-0_20

Vander Zanden, M.J. (2005). The success of animal invaders. *Proc Natl Acad Sci USA, 102,* 7055–7056. https://doi.org/10.1073/pnas.0502549102

Verbeek, P.-P. (2005). *What things do: Philosophical reflections on technology, agency, and design.* Pennsylvania State University Press. https://doi.org/10.1515/9780271033228

Verbeek, P.-P. (2006). Materializing morality: Design ethics and technological mediation. *Science, Technology & Human Values, 31*(3), 361–380. https://doi.org/10.1177/0162243905285847

Verbeek, P.-P. (2011). *Moralizing technology: understanding and designing the morality of things.* The University of Chicago Press. https://doi.org/10.7208/chicago/9780226852904.001.0001

Viacorka, F. [@franaviacorka] (2022, April 5). *Ukrainians are locating their devices on the territory of the Homiel region, Belarus, where part of the Russian army retreated.* [Post]. X. https://twitter.com/franakviacorka/status/1511297092836278283

Wang, Y. & Shi, Q. (2022). The impact of municipal solid waste sorting policy on air pollution: Evidence from Shanghai, China. *PLoS ONE, 17*(11). https://doi.org/10.1371/journal.pone.0277035

Wasley, A. (2018, July 5). *Two amputations a week: the cost of working in a US meat plant.* The Guardian. https://www.theguardian.com/environment/2018/jul/05/amputations-serious-injuries-us-meat-industry-plant

Watson, J. & Crick, F. (1953). Molecular structure of nucleic acids: a structure for deoxyribose nucleic acid. *Nature, 171,* 737–738. https://doi.org/10.1038/171737a0

Waymo. (n.d.). *The world's most experienced driver.* Waymo. https://waymo.com/

Webster, C. & Glasze, G. (2002). The global spread of gated communities. *Environment and Planning B: Planning and Design, 29*(3). 315-320. https://www.researchgate.net/publication/23541307_The_Global_Spread_of_Gated_Communities. https://doi.org/10.1068/b12926

Wellcome. (2003, March 30). *Francis Crick's controversial archive on first public display.* Wellcome. https://wellcome.org/press-release/francis-cricks-controversial-archive-first-public-display

NEOM. (n.d.). *What is NEOM?* NEOM. https://www.neom.com/en-us/about

Williams, E. E. & DeMello, M. (2007). *Why animals matter: the case for animal protection.* Prometheus Books.

Winner L. (1986). *The whale and the reactor: a search for limits in an age of high technology.* University of Chicago Press.

Winner, L. (2014). Technologies as forms of life. In Sandler, R.L. (eds) *Ethics and emerging technologies.* Palgrave Macmillan, London. https://doi.org/10.1057/9781137349088_4

World Association of Zoos and Aquariums (2016). Future of zoo and aquarium design. *WAZA Magazine,* 17. https://www.waza.org/wp-content/uploads/2019/02/Gusset__Chin_2016.pdf

Wray, B. (2017). *Rise of the Necrofauna: The Science, Ethics, and Risks of De-Extinction.* US: Greystone Books.

Youth v. Gov. (2023). *Historic climate trial: Held v. State of Montana.* https://www.youthvgov.org/held-v-montana

Zink, T., & Geyer, R. (2016). There is no such thing as a green product. *Stanford Social Innovation Review, 14*(2), 26–31. https://doi.org/10.48558/G401-WQ16

Wroclawski, D. & Bradley, M. (2024, March 12). *The ultimate guide to recycling K-Cups, Nespresso Capsules, and other coffee pods.* Consumer Reports. https://www.consumerreports.org/home-garden/recycling/recycle-k-cups-nespresso-capsules-coffee-pods-a3276889649/#:~:text=But%20the%20best%20solution%20to,preground%20or%20whole%2Dbean%20coffee

Zinsser, W. (2006). *On Writing Well (30th ed.).* HarperCollins.

Index

A

B

C

D

E

F

G

H

I

J

K

L

M

T

U

V

W

Z

www.ingramcontent.com/pod-product-compliance
Lightning Source LLC
LaVergne TN
LVHW020414220926
848484LV00018B/2568/J